Big Data Management

The big data paradigm presents a number of challenges for university curricula on big data or data science related topics. On the one hand, new research, tools and technologies are currently being developed to harness the increasingly large quantities of data being generated within our society. On the other, big data curricula at universities are still based on the computer science knowledge systems established in the 1960s and 70s. The gap between the theories and applications is becoming larger, as a result of which current education programs cannot meet the industry's demands for big data talents.

This series aims to refresh and complement the theory and knowledge framework for data management and analytics, reflect the latest research and applications in big data, and highlight key computational tools and techniques currently in development. Its goal is to publish a broad range of textbooks, research monographs, and edited volumes that will:

- Present a systematic and comprehensive knowledge structure for big data and data science research and education
- Supply lectures on big data and data science education with timely and practical reference materials to be used in courses
- Provide introductory and advanced instructional and reference material for students and professionals in computational science and big data
- Familiarize researchers with the latest discoveries and resources they need to advance the field
- Offer assistance to interdisciplinary researchers and practitioners seeking to learn more about big data

The scope of the series includes, but is not limited to, titles in the areas of database management, data mining, data analytics, search engines, data integration, NLP, knowledge graphs, information retrieval, social networks, etc. Other relevant topics will also be considered.

Xiang Zhao • Weixin Zeng • Jiuyang Tang

Entity Alignment

Concepts, Recent Advances and Novel Approaches

 Springer

Xiang Zhao
Laboratory for Big Data and Decision
National University of Defense Technology
Changsha, Hunan, China

Weixin Zeng
Laboratory for Big Data and Decision
National University of Defense Technology
Changsha, Hunan, China

Jiuyang Tang
Laboratory for Big Data and Decision
National University of Defense Technology
Changsha, Hunan, China

ISSN 2522-0179   ISSN 2522-0187   (electronic)
Big Data Management
ISBN 978-981-99-4252-7   ISBN 978-981-99-4250-3   (eBook)
https://doi.org/10.1007/978-981-99-4250-3

This work was partially supported by NSFC under grant Nos. 61872446, 62272469 and 71971212.

This Springer imprint is published by the registered company Springer Nature Singapore Pte Ltd.
The registered company address is: 152 Beach Road, #21-01/04 Gateway East, Singapore 189721, Singapore

Paper in this product is recyclable.

Preface

Background

Knowledge fusion is an important stage of knowledge management, which connects, combines and updates the knowledge from different resources. With the advent of the era of big data, knowledge graph (KG), as an effective means to extract structured knowledge from massive unstructured data and stocked structured data, becomes an essential part of knowledge management. KGs that are constructed by data-driven techniques usually come from different sources and have low coverage. Hence, it calls for establishing the connections among these individually constructed KGs using knowledge fusion techniques, which can thus achieve the augmentation and update of KGs. During the aforementioned process, entity alignment (EA) plays a crucial role. It aims to detect the equivalent entities in different KGs and connect heterogeneous KGs using these entities as anchors, which lays the foundation for the subsequent knowledge unification and update process. Currently, with the advancement of deep learning techniques, representation learning-based EA methods have become the mainstream approach.

Recent years have witnessed a rapid increase in the number of entity alignment frameworks, while the relationships among them remain unclear. This book aims to fill that gap by elaborating the concept and categorization of entity alignment, reviewing recent advances in entity alignment approaches, and introducing novel scenarios and corresponding solutions. Specifically, the book includes comprehensive evaluations and detailed analyses of state-of-the-art entity alignment approaches and strives to provide a clear picture of the strengths and weaknesses of the currently available solutions, so as to inspire follow-up research. In addition, it identifies novel entity alignment scenarios and explores the issues of large-scale data, long-tail knowledge, scarce supervision signals, lack of labeled data, and multimodal knowledge, offering potential directions for future research. The book offers a valuable reference guide for junior researchers, covering the latest advances in entity alignment, and a valuable asset for senior researchers, sharing novel entity alignment scenarios and their solutions. Accordingly, it will appeal to a broad audience in the fields of knowledge bases, database management, artificial intelligence, and big data.

Content Organization

The book consists of nine chapters, which can be divided into three parts.

Part I presents the background and overview of entity alignment and then discusses the state-of-the-art entity alignment solutions. Specifically, Chap. 1 presents a brief introduction to the entity alignment task. Besides, it also introduces some works that are closely related to entity alignment, as well as frequently used datasets. Chapter 2 conducts a comprehensive evaluation and detailed analysis of state-of-the-art EA approaches.

Part II introduces recent advances of entity alignment approaches, including progresses on representation learning (cf. Chap. 3) and alignment inference (cf. Chap. 4).

Part III introduces novel scenarios of entity alignment and corresponding solutions, including large-scale data, long-tail knowledge, scarce supervision signals, lack of labeled data, and multimodal knowledge, offering potential directions for future research.

Chapter 5 targets at entity alignment at scale, and puts forward a novel solution that can manage large-scale KG pairs and meanwhile achieve promising alignment performance.

Chapter 6 identifies the deficiency of existing EA methods in aligning long-tail entities, and approaches the limit by introducing a complementary signal from entity names in the form of concatenated power mean word embeddings and conceiving an effective way via degree-aware co-attention mechanism to dynamically fuse name and structural signals.

Chapter 7 tackles EA with scarce supervision. It puts forward a reinforced active entity alignment framework to select the entities to be manually labeled with the aim of enhancing alignment performance with minimal labeling efforts.

Chapter 8 identifies the deficiencies of existing EA methods, i.e., requiring labeled data and working under the closed-domain setting, and introduces an unsupervised EA framework to deal with unmatchable entities.

Chapter 9 introduces a novel multi-modal entity alignment strategy, i.e., hyperbolic multi-modal entity alignment, which extends the Euclidean representation to hyperboloid manifold.

Changsha, China   Xiang Zhao
May 2023   Weixin Zeng
                                                                         Jiuyang Tang

Contents

Part I
Concept and Categorization

Chapter 1
Introduction to Entity Alignment

Abstract In this section, we provide a concise overview of the entity alignment task and also discuss other related tasks that have a close connection to entity alignment.

1.1 Background

In the past few years, there has been a significant increase in the use and development of KGs and their various applications. These KGs are designed to store world knowledge, represented as triples (i.e., <entity, relation, entity>) consisting of entities, relations, and other entities, with each entity referring to a distinct real-world object, and each relation representing a connection between those objects. Since these entities serve as the foundation for the triples in a KG, the triples are inherently interconnected, creating a large and complex graph of knowledge. Currently, we have a large number of *general* KGs (e.g., DBpedia [1], YAGO [52], Google's Knowledge Vault [14]) and *domain*-specific KGs (e.g., medical [48] and scientific KGs [56]). KGs have been utilized to improve a wide range of downstream applications, including but not limited to keyword search [64], fact-checking [30], and question answering [12, 28].

A knowledge graph, denoted as $G = (E, R, T)$, is a graph that consists of three main components: a set of entities E, a set of relations R, and a set of triples T, where $T \subseteq E \times R \times E$ represents the directed edges in the graph. In the set of triples T, a single triple (h, r, t) represents a relationship between a head entity h and a tail entity t through a specific relation r. Each entity in the graph is identified by a unique identifier, such as http://dbpedia.org/resource/Spain in the case of DBpedia.

In practice, KGs are typically constructed from a single data source, making it difficult to achieve comprehensive coverage of a given domain [46]. To improve the completeness of a KG, one popular strategy is to integrate information from other KGs that may contain supplementary or complementary data. For instance, a general KG may only include basic information about a scientist, while scientific domain-specific KGs may have additional details like biographies and lists of publications.

© The Author(s) 2023

X. Zhao et al., *Entity Alignment*, Big Data Management,

https://doi.org/10.1007/978-981-99-4250-3_1

3

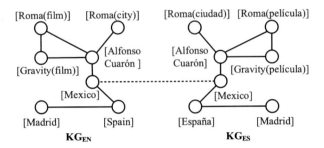

Fig. 1.1 An example of EA. The entity identifiers are placed in the square brackets. The prefixes of entity identifiers and the full relation identifiers are omitted for clarity; seed entity pairs are connected by dashed lines

To combine knowledge across multiple KGs, a crucial step is to align equivalent entities in different KGs, which is known as *entity alignment* (EA) [7, 25].[1]

Given a source KG $G_1 = (E_1, R_1, T_1)$, a target KG $G_2 = (E_2, R_2, T_2)$, and seed entity pairs (training set), i.e., $S = \{(u, v) \mid u \in E_1, v \in E_2, u \leftrightarrow v\}$, where $\leftrightarrow$ represents equivalence (i.e., u and v refer to the same real-world object), the task of EA can be defined as discovering the equivalent entity pairs in the test set.

Example Figure 1.1 shows a partial English KG (KG$_{EN}$) and a partial Spanish KG (KG$_{ES}$) concerning the director *Alfonso Cuarón*. Note that each entity in the KG has a unique identifier. For example, the movie "Roma" in the source KG is uniquely identified by Roma(film).[2] Given the seed entity pair, i.e., Mexico from KG$_{EN}$ and Mexico from KG$_{ES}$, EA aims to find the equivalent entity pairs in the test set, e.g., returning Roma(ciudad) in KG$_{ES}$ as the corresponding target entity to the source entity Roma(city) in KG$_{EN}$.

Broadly speaking, current entity alignment (EA) methods typically address the problem by assuming that equivalent entities in different KGs share similar local structures and applying representation learning techniques to embed entities as data points in a low-dimensional feature space. With effective entity embedding, the

---

[1] As where we are standing, EA can be deemed as a special case of entity resolution (ER), which recalls a pile of literature (to be discussed in Sect. 1.2). Thus, some ER methods (with minor adaptation to handle EA) are also involved in this book.

[2] The identifiers in some KGs are human-readable, e.g., those in Fig. 1.1, while some are incomprehensible, e.g., Freebase MIDs like /m/012rkqx.

pairwise dissimilarity of entities can be calculated as the distance between data points, allowing us to evaluate whether two entities are a match or not.[3]

1.2 Related Works

While the problem of EA was introduced a few years ago, the more generic version of the problem –identifying entity records referring to the same real-world entity from different data sources– has been investigated from various angles by different communities, under the names of entity resolution (ER) [15, 18, 45], entity matching [13, 42], record linkage [8, 34], deduplication [16], instance/ontology matching [20, 35, 49–51], link discovery [43, 44], and entity linking/entity disambiguation [11, 29]. Next, we describe the related work and the scope of this book.

1.2.1 Entity Linking

The process of entity linking (EL) or entity disambiguation is the act of recognizing entity mentions in natural language text and linking them to the corresponding entities in a given reference catalog, which is usually a knowledge graph. This process involves identifying which entity a particular mention in the text refers to. For example, if given the word "Rome," the task would be to determine if it refers to the city in Italy, a movie, or another entity and then link it to the right entity in the reference catalog. Prior studies in EL [21, 22, 29, 36, 68] have used various sources of information to disambiguate entity mentions, including surrounding words, prior probabilities of certain target entities, already disambiguated entity mentions, and background knowledge from sources such as Wikipedia. However, much of this information is not available in scenarios where aligning KGs is required, such as entity embeddings or the prior distribution of entity linking given a mention. Moreover, EL is concerned with mapping natural language text to a KG, while this research investigates the mapping of entities between two KGs.

1.2.2 Entity Resolution

Entity resolution, which is also referred to as entity matching, deduplication, or record linkage, assumes that the input is *relational data*, and each data object usually has a large amount of textual information described in multiple attributes. Therefore,

---

[3] Throughout the rest of this article, we may use the terms "align" and "match" interchangeably with the same meaning.

various similarity or distance functions are used in entity resolution to measure the similarity between two objects. These functions include Jaro-Winkler distance for comparing names and numerical distance for comparing dates. Based on the similarity measure, both rule-based and machine learning-based methods can be employed to classify two objects as either matching or non-matching [9].

To clarify further, in ER tasks, the attributes of data objects are first aligned, which can be done manually or automatically. Then, the similarity or distance functions are used to calculate the similarities between corresponding attribute values of the two objects. Finally, the similarity scores between the aligned attributes are combined or aggregated to determine the overall similarity between the two objects. This process allows rule-based or machine learning-based methods to classify pairs of objects as either matching or non-matching, based on the computed similarity scores [32, 45].

1.2.3 Entity Resolution on KGs

Certain methods for ER are created with the purpose of managing KGs and focus solely on binary connections, or data shaped like a graph. These methods are sometimes called instance/ontology matching approaches [49, 50]. The graph-shaped data comes with its own challenges: (1) Entities in graph-shaped data often lack detailed textual descriptions and may only be represented by their name, with a minimal amount of accompanying information. (2) Unlike classical databases, which assume that all fields of a record are present, KGs are built on the Open World Assumption, where the absence of certain attributes of an entity in the KG does not necessarily mean that they do not exist in reality. This fundamental difference sets KGs apart from traditional databases. (3) KGs have their own set of predefined semantics. At a basic level, these can take the form of a taxonomy of classes. In more complex cases, KGs can be endowed with an ontology of logical axioms.

In the past 20 years, various techniques have been developed to address the specific challenges of KGs, particularly in the context of the Semantic Web and the Linked Open Data cloud [26]. These techniques can be categorized along several different dimensions:

- **Scope.** Several techniques have been developed for aligning KGs along different dimensions. For example, some approaches aim to align the entities in two different KGs, while others focus on aligning the relationship names, or schema, between KGs. Additionally, some methods aim to align the class taxonomies of two KGs, and a few techniques achieve all three tasks at once. In this particular book, however, the focus is on the first task, which is aligning entities in KGs.

- **Background knowledge.** Certain techniques rely on an ontology (T-box) as background information, particularly those that participate in the Ontology Alignment Evaluation Initiative (OAEI).[4] However, in this specific book, the focus is on techniques that do not require such prior knowledge and can operate without an ontology.
- **Training.** Some techniques for aligning knowledge graphs are unsupervised and operate directly on input data without any need for training data or a training phase. Examples of such methods include PARIS [51] and SiGMa [35]. On the other hand, other approaches involve learning mappings between entities based on predefined seeds. This particular book, however, focuses on the latter class of approaches.

Most of the supervised or semi-supervised approaches for entity alignment utilize recent advances in deep learning [23]. These approaches primarily rely on graph representation learning techniques to model the structure of knowledge graphs and generate entity embeddings for alignment. To refer to the supervised or semi-supervised approaches, we use the term "entity alignment (EA) approaches," which is also the main focus of this study. However, in the next chapter, we include PARIS [51] for comparison as a representative of the unsupervised approaches. We also include AgreementMakerLight (AML) [17] as a representative of unsupervised systems that use background knowledge. For the other systems, we refer the reader to other surveys [9, 33, 41, 43].

In addition, since EA pursues the same goal as ER, it can be deemed a special but nontrivial case of ER. In this light, general ER approaches can be adapted to the problem of EA, and we include representative ER methods for comparison (to be detailed in Chap. 2).

Existing Benchmarks Several synthetic datasets, such as DBP15K and DWY100K, were created using the inter-language and reference links already present in DBpedia to assess the effectiveness of EA methods. Chapter 2 contains more extensive statistical information about these datasets.

Notably, the Ontology Alignment Evaluation Initiative (OAEI) promoted the knowledge graph track.[5] Existing benchmarks for EA only provide instance-level information, while the KGs in these datasets include both schema and instance information. This can create an unfair evaluation of current EA approaches that do not consider the availability of ontology information. Hence, they are not presented in this book.

---

[4] http://oaei.ontologymatching.org/.

[5] http://oaei.ontologymatching.org/2019/knowledgegraph.

1.3   Evaluation Settings

This section provides an introduction to the evaluation settings that are commonly used for the EA task.

Datasets Three datasets are commonly used and are representative, including the following:

- DBP15K [53]. This particular dataset comprises three pairs of multilingual KGs that were extracted from DBpedia. These pairs include English to Chinese (DBP15K$_{ZH-EN}$), English to Japanese (DBP15K$_{JA-EN}$), and English to French (DBP15K$_{FR-EN}$). Each of these KG pairs is made up of 15,000 inter-language links, which serve as gold standards.
- DWY100K [54]. The dataset consists of two pairs of mono-lingual knowledge graphs, namely, DWY100K$_{DBP-WD}$ and DWY100K$_{DBP-YG}$. These pairs were extracted from DBpedia, Wikidata, and YAGO 3, and each one contains 100,000 pairs of entities. The extraction process is similar to that of DBP15K, except that the inter-language links have been replaced with reference links that connect these knowledge graphs.
- SRPRS. According to Guo et al. [24], the KGs in previous EA datasets, such as DBP15K and DWY100K, are overly dense and do not accurately reflect the degree distributions observed in real-life KGs. In response to this issue, Guo et al. [24] developed a new EA benchmark that uses reference links in DBpedia to establish KGs with degree distributions that better reflect real-life situations. The resulting evaluation benchmark includes both cross-lingual (SRPRS$_{EN-FR}$, SRPRS$_{EN-DE}$) and mono-lingual KG pairs (SRPRS$_{DBP-WD}$, SRPRS$_{DBP-YG}$), where EN, FR, DE, DBP, WD, and YG represent DBpedia (English), DBpedia (French), DBpedia (German), DBpedia, Wikidata, and YAGO 3, respectively. Each KG pair is comprised of 15,000 pairs of entities.

Table 1.1 provides a summary of the datasets used in this study. Each KG pair includes relational triples, cross-KG entity pairs (30% of which are seed entity pairs and used for training), and attribute triples. The cross-KG entity pairs serve as gold standards.

Degree Distribution Figure 1.2 presents the degree distributions of entities in the datasets, which provides insights into the characteristics of these datasets. The *degree* of an entity is defined as the number of triples in which the entity is involved. Entities with higher degrees tend to have richer neighboring structures. The degree distributions of the different KG pairs in each dataset are very similar. Thus, for brevity, we present only one KG pair's degree distribution in Fig. 1.2.

The sub-figures in series (a) correspond to the DBP15K dataset. As shown, entities with a degree of 1 comprise the largest proportion, while the number of entities generally decreases with increasing degree values, with some fluctuations.

Table 1.1 Statistics of EA benchmarks and our constructed dataset

Name	KG pair	#Triples	#Entities	#Relations	#Align.
DBP15K$_{ZH-EN}$	DBpedia(Chinese)	70,414	19,388	1,701	15,000
	DBpedia (English)	95,142	19,572	1,323	
DBP15K$_{JA-EN}$	DBpedia(Japanese)	77,214	19,814	1,299	15,000
	DBpedia (English)	93,484	19,780	1,153	
DBP15K$_{FR-EN}$	DBpedia(French)	105,998	19,661	903	15,000
	DBpedia (English)	115,722	19,993	1,208	
SRPRS$_{EN-FR}$	DBpedia(English)	36,508	15,000	221	15,000
	DBpedia (French)	33,532	15,000	177	
SRPRS$_{EN-DE}$	DBpedia(English)	38,281	15,000	222	15,000
	DBpedia (German)	37,069	15,000	120	
SRPRS$_{DBP-WD}$	DBpedia	38,421	15,000	253	15,000
	Wikidata	40,159	15,000	144	
SRPRS$_{DBP-YG}$	DBpedia	33,571	15,000	223	15,000
	YAGO3	34,660	15,000	30	
DWY100K$_{DBP-WD}$	DBpedia	463,294	100,000	330	100,000
	Wikidata	448,774	100,000	220	
DWY100K$_{DBP-YG}$	DBpedia	428,952	100,000	302	100,000
	YAGO3	502,563	100,000	31	

It is worth noting that the coverage curve approximates a straight line, as the number of entities changes only slightly when the degree increases from 2 to 10.

The (b) set of figures is related to DWY100K. This dataset has a distinct structure from (a), as there are no entities with a degree of 1 or 2. Additionally, the number of entities reaches its highest point at degree 4 and then decreases as the entity degree increases.

The (c) set of figures is related to SRPRS. It is clear that the degree distribution of entities in this dataset is more realistic, with entities of lower degrees making up a larger proportion. This is due to its well-thought-out sampling approach. Additionally, the (d) set of figures corresponds to the dataset we created, which will be discussed in Chap. 2.

Evaluation Metrics Most existing EA solutions use Hits@k ($k = 1, 10$) and mean reciprocal rank (MRR) as their evaluation metrics. The target entities are arranged in order of increasing distance scores from the source entity when making a prediction. The Hits@k metric shows the proportion of correctly aligned entities among the k nearest target entities. Hits@1 is the most significant measure of the accuracy of the alignment results.

MRR denotes the average of the reciprocal ranks of the ground truths. Note that higher Hits@k and MRR indicate better performance. Unless otherwise specified, the results of Hits@k are represented in percentages.

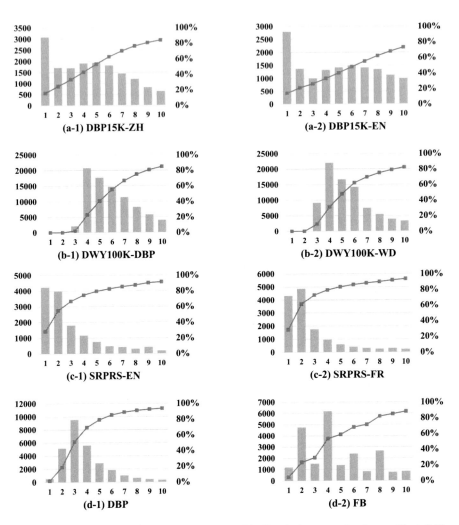

Fig. 1.2 Degree distributions on different datasets. The X-axis denotes entity degree. The left Y-axis represents the number of entities (corresponding to bars), while the right Y-axis represents the percentage of entities with a degree lower than a given x value (corresponding to lines)

References

1. S. Auer, C. Bizer, G. Kobilarov, J. Lehmann, R. Cyganiak, and Z. G. Ives. Dbpedia: A nucleus for a web of open data. In *ISWC*, pages 722–735, 2007.
2. P. Bojanowski, E. Grave, A. Joulin, and T. Mikolov. Enriching word vectors with subword information. *Transactions of the Association for Computational Linguistics*, 5:135–146, 2017.
3. K. D. Bollacker, C. Evans, P. Paritosh, T. Sturge, and J. Taylor. Freebase: a collaboratively created graph database for structuring human knowledge. In *SIGMOD*, pages 1247–1250, 2008.
4. A. Bordes, N. Usunier, A. García-Durán, J. Weston, and O. Yakhnenko. Translating embeddings for modeling multi-relational data. In *NIPS*, pages 2787–2795, 2013.

5. Y. Cao, Z. Liu, C. Li, Z. Liu, J. Li, and T. Chua. Multi-channel graph neural network for entity alignment. In *ACL*, pages 1452–1461, 2019.
6. M. Chen, Y. Tian, K. Chang, S. Skiena, and C. Zaniolo. Co-training embeddings of knowledge graphs and entity descriptions for cross-lingual entity alignment. In *IJCAI*, pages 3998–4004, 2018.
7. M. Chen, Y. Tian, M. Yang, and C. Zaniolo. Multilingual knowledge graph embeddings for cross-lingual knowledge alignment. In *IJCAI*, pages 1511–1517, 2017.
8. P. Christen. A survey of indexing techniques for scalable record linkage and deduplication. *IEEE Trans. Knowl. Data Eng.*, 24(9):1537–1555, 2012.
9. V. Christophides, V. Efthymiou, T. Palpanas, G. Papadakis, and K. Stefanidis. End-to-end entity resolution for big data: A survey. *CoRR*, abs/1905.06397, 2019.
10. A. Conneau, G. Lample, M. Ranzato, L. Denoyer, and H. Jégou. Word translation without parallel data. *arXiv preprint arXiv:1710.04087*, 2017.
11. S. Cucerzan. Large-scale named entity disambiguation based on wikipedia data. In *EMNLP-CoNLL*, pages 708–716, 2007.
12. W. Cui, Y. Xiao, H. Wang, Y. Song, S. Hwang, and W. Wang. KBQA: learning question answering over QA corpora and knowledge bases. *PVLDB*, 10(5):565–576, 2017.
13. S. Das, P. S. G. C., A. Doan, J. F. Naughton, G. Krishnan, R. Deep, E. Arcaute, V. Raghavendra, and Y. Park. Falcon: Scaling up hands-off crowdsourced entity matching to build cloud services. In *SIGMOD*, pages 1431–1446, 2017.
14. X. Dong, E. Gabrilovich, G. Heitz, W. Horn, N. Lao, K. Murphy, T. Strohmann, S. Sun, and W. Zhang. Knowledge vault: a web-scale approach to probabilistic knowledge fusion. In *KDD*, pages 601–610, 2014.
15. M. Ebraheem, S. Thirumuruganathan, S. R. Joty, M. Ouzzani, and N. Tang. Distributed representations of tuples for entity resolution. *PVLDB*, 11(11):1454–1467, 2018.
16. A. K. Elmagarmid, P. G. Ipeirotis, and V. S. Verykios. Duplicate record detection: A survey. *IEEE Trans. Knowl. Data Eng.*, 19(1):1–16, 2007.
17. D. Faria, C. Pesquita, E. Santos, I. F. Cruz, and F. M. Couto. Agreementmakerlight 2.0: Towards efficient large-scale ontology matching. In M. Horridge, M. Rospocher, and J. van Ossenbruggen, editors, *ISWC*, volume 1272 of *CEUR Workshop Proceedings*, pages 457–460. CEUR-WS.org, 2014.
18. C. Fu, X. Han, L. Sun, B. Chen, W. Zhang, S. Wu, and H. Kong. End-to-end multi-perspective matching for entity resolution. In *IJCAI*, pages 4961–4967, 2019.
19. L. Galárraga, C. Teflioudi, K. Hose, and F. M. Suchanek. Fast rule mining in ontological knowledge bases with AMIE+. *VLDB J.*, 24(6):707–730, 2015.
20. L. A. Galárraga, N. Preda, and F. M. Suchanek. Mining rules to align knowledge bases. In *AKBC@CIKM*, pages 43–48, 2013.
21. O.-E. Ganea and T. Hofmann. Deep joint entity disambiguation with local neural attention. In *EMNLP*, pages 2619–2629, Sept. 2017.
22. A. Globerson, N. Lazic, S. Chakrabarti, A. Subramanya, M. Ringgaard, and F. Pereira. Collective entity resolution with multi-focal attention. In *ACL*, pages 621–631, Aug. 2016.
23. I. J. Goodfellow, Y. Bengio, and A. C. Courville. *Deep Learning*. Adaptive computation and machine learning. MIT Press, 2016.
24. L. Guo, Z. Sun, and W. Hu. Learning to exploit long-term relational dependencies in knowledge graphs. In *ICML*, pages 2505–2514, 2019.
25. Y. Hao, Y. Zhang, S. He, K. Liu, and J. Zhao. A joint embedding method for entity alignment of knowledge bases. In *CCKS*, pages 3–14, 2016.
26. T. Heath and C. Bizer. *Linked Data: Evolving the Web into a Global Data Space*. Synthesis Lectures on the Semantic Web. Morgan & Claypool Publishers, 2011.
27. S. Hertling and H. Paulheim. The knowledge graph track at OAEI - gold standards, baselines, and the golden hammer bias. In A. Harth, S. Kirrane, A. N. Ngomo, H. Paulheim, A. Rula, A. L. Gentile, P. Haase, and M. Cochez, editors, *ESWC*, volume 12123 of *Lecture Notes in Computer Science*, pages 343–359. Springer, 2020.

28. B. Hixon, P. Clark, and H. Hajishirzi. Learning knowledge graphs for question answering through conversational dialog. In *NAACL*, pages 851–861, 2015.
29. J. Hoffart, M. A. Yosef, I. Bordino, H. Fürstenau, M. Pinkal, M. Spaniol, B. Taneva, S. Thater, and G. Weikum. Robust disambiguation of named entities in text. In *EMNLP*, pages 782–792, 2011.
30. V. Huynh and P. Papotti. Buckle: Evaluating fact checking algorithms built on knowledge bases. *PVLDB*, 12(12):1798–1801, 2019.
31. T. N. Kipf and M. Welling. Semi-supervised classification with graph convolutional networks. *CoRR*, abs/1609.02907, 2016.
32. P. Konda, S. Das, P. S. G. C., A. Doan, A. Ardalan, J. R. Ballard, H. Li, F. Panahi, H. Zhang, J. F. Naughton, S. Prasad, G. Krishnan, R. Deep, and V. Raghavendra. Magellan: Toward building entity matching management systems. *PVLDB*, 9(12):1197–1208, 2016.
33. H. Köpcke and E. Rahm. Frameworks for entity matching: A comparison. *Data Knowl. Eng.*, 69(2):197–210, 2010.
34. N. Koudas, S. Sarawagi, and D. Srivastava. Record linkage: similarity measures and algorithms. In *SIGMOD*, pages 802–803, 2006.
35. S. Lacoste-Julien, K. Palla, A. Davies, G. Kasneci, T. Graepel, and Z. Ghahramani. Sigma: simple greedy matching for aligning large knowledge bases. In *KDD*, pages 572–580, 2013.
36. P. Le and I. Titov. Improving entity linking by modeling latent relations between mentions. In *ACL*, pages 1595–1604, 2018.
37. V. I. Levenshtein. Binary codes capable of correcting deletions, insertions, and reversals. In *Soviet physics doklady*, volume 10, pages 707–710, 1966.
38. C. Li, Y. Cao, L. Hou, J. Shi, J. Li, and T.-S. Chua. Semi-supervised entity alignment via joint knowledge embedding model and cross-graph model. In *EMNLP*, pages 2723–2732, 2019.
39. Y. Liu, H. Li, A. García-Durán, M. Niepert, D. Oñoro-Rubio, and D. S. Rosenblum. MMKG: multi-modal knowledge graphs. In P. Hitzler, M. Fernández, K. Janowicz, A. Zaveri, A. J. G. Gray, V. López, A. Haller, and K. Hammar, editors, *ESWC*, volume 11503 of *Lecture Notes in Computer Science*, pages 459–474. Springer, 2019.
40. F. Monti, O. Shchur, A. Bojchevski, O. Litany, S. Günnemann, and M. M. Bronstein. Dual-primal graph convolutional networks. *CoRR*, abs/1806.00770, 2018.
41. M. Mountantonakis and Y. Tzitzikas. Large-scale semantic integration of linked data: A survey. *ACM Comput. Surv.*, 52(5):103:1–103:40, 2019.
42. S. Mudgal, H. Li, T. Rekatsinas, A. Doan, Y. Park, G. Krishnan, R. Deep, E. Arcaute, and V. Raghavendra. Deep learning for entity matching: A design space exploration. In *SIGMOD*, pages 19–34, 2018.
43. M. Nentwig, M. Hartung, A. N. Ngomo, and E. Rahm. A survey of current link discovery frameworks. *Semantic Web*, 8(3):419–436, 2017.
44. A. N. Ngomo and S. Auer. LIMES - A time-efficient approach for large-scale link discovery on the web of data. In *IJCAI*, pages 2312–2317, 2011.
45. H. Nie, X. Han, B. He, L. Sun, B. Chen, W. Zhang, S. Wu, and H. Kong. Deep sequence-to-sequence entity matching for heterogeneous entity resolution. In *CIKM*, pages 629–638, 2019.
46. H. Paulheim. Knowledge graph refinement: A survey of approaches and evaluation methods. *Semantic Web*, 8(3):489–508, 2017.
47. V. Rastogi, N. N. Dalvi, and M. N. Garofalakis. Large-scale collective entity matching. *PVLDB*, 4(4):208–218, 2011.
48. M. Rotmensch, Y. Halpern, A. Tlimat, S. Horng, and D. Sontag. Learning a health knowledge graph from electronic medical records. *Scientific Reports*, 7, 12 2017.
49. C. Shao, L. Hu, J. Li, Z. Wang, T. L. Chung, and J. Xia. Rimom-im: A novel iterative framework for instance matching. *J. Comput. Sci. Technol.*, 31(1):185–197, 2016.
50. P. Shvaiko and J. Euzenat. Ontology matching: State of the art and future challenges. *IEEE Trans. Knowl. Data Eng.*, 25(1):158–176, 2013.
51. F. M. Suchanek, S. Abiteboul, and P. Senellart. PARIS: probabilistic alignment of relations, instances, and schema. *PVLDB*, 5(3):157–168, 2011.

52. F. M. Suchanek, G. Kasneci, and G. Weikum. Yago: a core of semantic knowledge. In *WWW*, pages 697–706, 2007.
53. Z. Sun, W. Hu, and C. Li. Cross-lingual entity alignment via joint attribute-preserving embedding. In *ISWC*, pages 628–644, 2017.
54. Z. Sun, W. Hu, Q. Zhang, and Y. Qu. Bootstrapping entity alignment with knowledge graph embedding. In *IJCAI*, pages 4396–4402, 2018.
55. Z. Sun, J. Huang, W. Hu, M. Chen, L. Guo, and Y. Qu. Transedge: Translating relation-contextualized embeddings for knowledge graphs. In *ISWC*, pages 612–629, 2019.
56. J. Tang, J. Zhang, L. Yao, J. Li, L. Zhang, and Z. Su. Arnetminer: extraction and mining of academic social networks. In *SIGKDD*, pages 990–998. ACM, 2008.
57. B. D. Trisedya, J. Qi, and R. Zhang. Entity alignment between knowledge graphs using attribute embeddings. In *AAAI*, pages 297–304, 2019.
58. A. Vaswani, N. Shazeer, N. Parmar, J. Uszkoreit, L. Jones, A. N. Gomez, L. Kaiser, and I. Polosukhin. Attention is all you need. In *NIPS*, pages 5998–6008, 2017.
59. Z. Wang, Q. Lv, X. Lan, and Y. Zhang. Cross-lingual knowledge graph alignment via graph convolutional networks. In *EMNLP*, pages 349–357, 2018.
60. Y. Wu, X. Liu, Y. Feng, Z. Wang, R. Yan, and D. Zhao. Relation-aware entity alignment for heterogeneous knowledge graphs. In *IJCAI*, pages 5278–5284, 2019.
61. Y. Wu, X. Liu, Y. Feng, Z. Wang, and D. Zhao. Jointly learning entity and relation representations for entity alignment. In *EMNLP*, pages 240–249, 2019.
62. K. Xu, L. Wang, M. Yu, Y. Feng, Y. Song, Z. Wang, and D. Yu. Cross-lingual knowledge graph alignment via graph matching neural network. In *ACL*, pages 3156–3161, 2019.
63. H.-W. Yang, Y. Zou, P. Shi, W. Lu, J. Lin, and S. Xu. Aligning cross-lingual entities with multi-aspect information. In *EMNLP*, pages 4422–4432, 2019.
64. Y. Yang, D. Agrawal, H. V. Jagadish, A. K. H. Tung, and S. Wu. An efficient parallel keyword search engine on knowledge graphs. In *ICDE*, pages 338–349, 2019.
65. W. Zeng, X. Zhao, J. Tang, and X. Lin. Collective entity alignment via adaptive features. In *ICDE*, pages 1870–1873. IEEE, 2020.
66. W. Zeng, X. Zhao, W. Wang, J. Tang, and Z. Tan. Degree-aware alignment for entities in tail. In *SIGIR*, pages 811–820. ACM, 2020.
67. Q. Zhang, Z. Sun, W. Hu, M. Chen, L. Guo, and Y. Qu. Multi-view knowledge graph embedding for entity alignment. In *IJCAI*, pages 5429–5435, 2019.
68. X. Zhou, Y. Miao, W. Wang, and J. Qin. A recurrent model for collective entity linking with adaptive features. In *AAAI*, pages 329–336. AAAI Press, 2020.
69. H. Zhu, R. Xie, Z. Liu, and M. Sun. Iterative entity alignment via joint knowledge embeddings. In *IJCAI*, pages 4258–4264, 2017.
70. Q. Zhu, X. Zhou, J. Wu, J. Tan, and L. Guo. Neighborhood-aware attentional representation for multilingual knowledge graphs. In *IJCAI*, pages 1943–1949, 2019.

Chapter 2
State-of-the-Art Approaches

Abstract This chapter performs a thorough assessment and meticulous examination of the most advanced EA techniques. Initially, we introduce a broad EA framework that covers all current methods and classify these methods into three main groups. Then, we carefully appraise these solutions on various scenarios, taking into account their efficacy, efficiency, and scalability. Lastly, we create a novel EA dataset that reflects the actual difficulties encountered in alignment, which prior literature mostly ignored. This chapter aims to offer a comprehensive understanding of the advantages and drawbacks of current EA methods, in order to encourage further high-quality research.

2.1 Introduction

In this chapter, we conduct an empirical evaluation of state-of-the-art EA approaches, which possesses the following characteristics:

Fair Comparison Within and Across Categories Most recent studies have limited themselves to comparing only a subset of methods [4, 11, 15, 23, 27–30, 33]. Moreover, different approaches follow different protocols: some use only the KG structure for alignment, while others incorporate additional information; some perform one-pass alignment of KGs, while others use an iterative (re-)training strategy. While the literature presents a direct comparison of these methods, which highlights their overall effectiveness, a more desirable and equitable approach would be to classify these methods into categories and then compare the outcomes within and across categories.

In this chapter, we incorporate most of the state-of-the-art methods to facilitate a comprehensive comparison, including the very recent approaches that have not been evaluated against other methods previously. We divide them into three groups and conduct a thorough analysis of both intra- and inter-group evaluations, enabling us to better position these methods and evaluate their effectiveness.

© The Author(s) 2023

X. Zhao et al., *Entity Alignment*, Big Data Management,

https://doi.org/10.1007/978-981-99-4250-3_2

Comprehensive Evaluation on Representative Datasets To assess the performance of EA systems, various datasets have been developed, which can be broadly classified into two categories: *cross-lingual* benchmarks, exemplified by DBP15K [21], and *mono-lingual* benchmarks, exemplified by DWY100K [22]. A recent study [11] highlights that the KGs in prior datasets are much *denser* than those in real-world scenarios, which led them to create the SRPRS dataset with entity degrees that follow a *normal* distribution. Despite the availability of multiple datasets, previous studies only report their results on one or two specific datasets, making it challenging to evaluate their efficacy across a wide range of potential scenarios, such as cross-lingual/mono-lingual, dense/normal, and large-scale/medium-scale KGs.

In light of this observation, this chapter performs a thorough experimental evaluation on all the prominent datasets, namely, DBP15K, DWY100K, and SRPRS, which together consist of nine pairs of knowledge graphs. The evaluation is conducted across various dimensions, including effectiveness, efficiency, and robustness.

New Dataset for Real-Life Challenges It has been noted that current EA datasets assume that each entity in the source KG has exactly one corresponding entity in the target KG, which is an unrealistic assumption. In reality, there are entities in one KG that may not have a corresponding entity in the other KG. For example, when aligning YAGO 4 and IMDB, only a small percentage (1%) of entities in YAGO 4 are related to movies, while the remaining 99% of entities in YAGO 4 do not have any corresponding entities in IMDB. These unmatchable entities would make the EA task more challenging.

Furthermore, we notice that the mono-lingual datasets currently available for EA evaluation assume that the entities in the different KGs share the same naming convention. Therefore, the baseline method that relies on comparing the string similarity between entity names can achieve perfect accuracy. However, this assumption is often not valid in real-life scenarios, where equivalent entities in different KGs may have dissimilar names, such as "America" and "USA" for the same entity. In addition, another challenge that is often overlooked in EA is that different entities in a KG might have the same name. This can make it difficult to determine whether an entity with the name "Paris" in the source KG refers to the same entity as one with the same name in the target KG, as they could potentially refer to different entities, such as the city in France and the city in Texas.

For these reasons, we believe that the current EA datasets do not fully capture the realistic challenges posed by unmatchable entities and ambiguous entity names. To address this issue, we introduce a new dataset that more closely mirrors these practical difficulties.

The main contributions of this chapter are the following:

- This chapter provides a comprehensive evaluation of state-of-the-art EA approaches. The evaluation includes: (1) Identifying the main components of existing EA approaches and proposing a general EA framework (2) Categorizing state-of-the-art approaches into three groups and conducting detailed intra- and inter-group evaluations to better understand their strengths and weaknesses (3) Examining these approaches in various scenarios, including cross-/mono-lingual alignment and alignment on dense/normal, large-/medium-scale data, to evaluate their *effectiveness*, *efficiency*, and *robustness*. The empirical results provide insights into the performance of each approach. This evaluation aims to provide a more systematic and comprehensive understanding of the current state of EA research.

- Through our study, we gained valuable experience and insights that allow us to identify the shortcomings of current EA datasets. To address these issues, we have created a new mono-lingual dataset that accurately reflects the real-life challenges of unmatchable entities and ambiguous entity names. We anticipate that this new dataset will provide a more effective benchmark for evaluating EA systems.

2.2   A General EA Framework

This section presents a general EA framework that is designed to include state-of-the-art EA approaches. Through a thorough analysis of current EA approaches, we identify four primary components, as shown in Fig. 2.1:

- **Embedding learning module.** This component is designed to train embeddings for entities, which can be broadly classified into two groups: KG representation-based models such as TransE [3] and graph neural network (GNN)-based models such as the graph convolutional network (GCN) [13].
- **Alignment module.** This component focuses on aligning the entity embeddings learned in the previous module across different KGs. The goal is to map these embeddings into a unified space. Margin-based loss is a common approach used in this module to ensure that the seed entity embeddings from different KGs are close to each other. Another approach used frequently is corpus fusion, which aligns KGs at the corpus level and directly embeds entities in different KGs into the same vector space.
- **Prediction module.** Once the unified embedding space is established, the next step is to predict the corresponding target entity for each source entity in the test set. One common approach is to use distance-based similarity measures such as cosine similarity, Manhattan distance, or Euclidean distance between entity

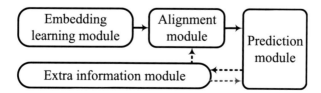

Fig. 2.1   A general EA framework

embeddings to calculate the similarity between entities. The target entity with the highest similarity (or lowest distance) is then selected as the counterpart.

- **Extra information module.** In addition to the basic modules, some EA approaches use additional information to improve their performance. One approach is bootstrapping, where confident alignment results are used as training data for subsequent alignment iterations. Another approach is to use multi-type literal information such as attributes, entity descriptions, and entity names to complement the KG structure. These additional sources of information are shown in Fig. 2.1 as blue dashed lines.

Example Further to the example in Chap. 1, we explain these modules. The *embedding learning module* generates embeddings for entities in KG_{EN} and KG_{ES}, respectively. Then the *alignment module* projects the entity embeddings into the same vector space, where the entity embeddings in KG_{EN} and KG_{ES} are directly comparable. Finally, using the unified embeddings, the *prediction module* aims to predict the equivalent target entity in KG_{ES} for each source entity in KG_{EN}. The *extra information module* leverages several techniques to improve the EA performance. Concretely, the bootstrapping strategy aims to include the confident EA pairs detected from a previous round, e.g., (Spain, España), into the training set for learning in the next round. Another approach is to use additional textual information to complement the entity embeddings for alignment.

We organize the state-of-the-art approaches based on each module of the EA framework and present them in Table 2.1. For a more detailed view of the approaches, readers can refer to the Appendix. Now, we will explain how each of these modules is implemented in various state-of-the-art approaches.

2.2.1 Embedding Learning Module

In this section, we will explain the techniques used in the embedding learning module, which utilize the KG structure to create embeddings for each entity.

Table 2.1 shows that the most commonly used models for this module are TransE [3] and GCN [13]. We will provide a brief overview of these fundamental models.

TransE The TransE model views relationships as translations that act on the lower-dimensional representations of entities. To clarify, when presented with a relational triple (h, r, t), TransE proposes that the embedded representation of the tail entity t should be similar to the embedded representation of the head entity h plus the

Table 2.1 A summary of the EA approaches involved in this study

Method	Embedding	Alignment	Extra info.	Prediction	C-L[a]	M-L	Group
MTransE [6]	TransE*[b]	Transition	✗	Euclidean distance	✓	✗	I
RSNs [11]	RSNs	Corpus fusion	✗	Cosine similarity	✓	✓	I
MuGNN [4]	GNN	Margin-based	✗	Cosine similarity	✓	✓	I
KECG [15]	GAT+TransE	Margin-based	✗	Euclidean distance	✓	✓	I
ITransE [34]	TransE	Transition	Bootstrapping	Manhattan distance	✗	✓	II
BootEA [22]	TransE*	Corpus fusion	Bootstrapping	Cosine similarity	✓	✓	II
NAEA [35]	TransE*	Corpus fusion	Bootstrapping	Cosine similarity	✓	✓	II
TransEdge [23]	TransEdge	Corpus fusion	Bootstrapping	Cosine similarity	✓	✓	II
JAPE [21]	TransE*	Attribute-refined	Attribute	Cosine similarity	✓	✗	III
GCN-Align [26]	GCN	Margin-based	Attribute	Manhattan distance	✓	✗	III
AttrE [24]	TransE	Attribute-refined	Attribute	Cosine similarity	✗	✓	III
KDCoE [5]	TransE	Transition	Entity description	Euclidean distance	✓	✗	III
HMAN [30]	GCN	Margin-based	Description, attribute	Manhattan distance	✓	✗	III
GM-Align [29]	GCN	Graph matching	Entity name	Matching probability	✓	✗	III
RDGCN [27]	DPGCNN	Margin-based	Entity name	Manhattan distance	✓	✗	III
HGCN [28]	GCN	Margin-based	Entity name	Manhattan distance	✓	✗	III
M-Greedy [33]	TransE*	Corpus fusion	Entity name, attribute	Cosine similarity	✗	✓	III
CEA [31]	GCN	Margin-based	Entity name	Cosine similarity	✓	✓	III

[a] **C-L** stands for **cross-lingual evaluation** and **M-L** stands for **mono-lingual evaluation**
[b] TransE* represents variants of the TransE model

embedded representation of the relationship r, or $\mathbf{h} + \mathbf{r} \approx \mathbf{t}$. By doing so, the model is able to maintain the structural information of the entities and produce close representations for entities that share similar neighbors in the embedding space.

GCN A type of convolutional network that processes graph-based data directly is known as the graph convolutional network (GCN). It creates embeddings for individual nodes by encoding information about the neighborhoods of those nodes. GCN takes as input feature vectors for each node in the KG, as well as a representative graph structure description in matrix form, such as an adjacency matrix. The output of the GCN is a new feature matrix. A typical GCN model consists of multiple stacked GCN layers, which allows it to capture a partial KG structure that extends several hops away from the entity being processed.

On top of these basic models, some methods make modifications. Regarding the TransE-based models, MTransE removes the negative triples during training, BootEA and NAEA replace the original margin-based loss function with a limit-based objective function, MuGNN uses the logistic loss to substitute for the margin-based loss, and JAPE designs a new loss function.

Concerning the GCN-based models, it has been observed that the GCN does not take into account the relations present in KGs. Therefore, as a solution, RDGCN employs the dual-primal graph convolutional neural network (DPGCNN) [17]. In contrast, MuGNN leverages an attention-based GNN model to assign varying weights to neighboring nodes. Additionally, KECG merges graph attention network (GAT) [25] and TransE to capture both the inner-graph structure and the inter-graph alignment information.

Several approaches have introduced new embedding models. For example, in RSNs, the authors contend that triple-level learning is inadequate for capturing long-term relational dependencies between entities and is insufficient for propagating semantic information among entities. Therefore, they propose using recurrent neural networks (RNNs) with residual learning to learn the long-term relational paths between entities.

Similarly, TransEdge devises a new energy function to measure the error of edge translation between entity embeddings for KG embedding learning. This method models edge embeddings using context compression and projection.

2.2.2 Alignment Module

In this subsection, we introduce the methods used for the alignment module, which aims to unify separated KG embeddings.

The prevailing approach in KG embedding learning is to use a margin-based loss function on top of the embedding learning module. This loss function requires that the distance between entities in *positive pairs* should be small, while the distance between entities in *negative pairs* should be large, with a margin between the distances of positive and negative pairs. The *positive pairs* refer to seed entity pairs, while *negative pairs* are generated by corrupting the positive pairs. This approach helps to merge the two separate KG embedding spaces into one vector space. Table 2.1 indicates that the majority of methods that use GNNs rely on a margin-based alignment model to merge the two KG embedding spaces. In contrast, in GM-Align, a matching framework is employed to maximize the matching probabilities of seed entity pairs, which achieves the alignment process.

Corpus fusion is another common approach, which involves using the seed entity pairs to connect the training corpora of two KGs. Some methods, such as BootEA and NAEA, generate new triples by swapping the entities in the seed entity pairs to align the embeddings in a unified space. Concretely, given an entity pair (u, v), the newly generated triples for G_1 are $T_1^{new} = \{(v, r, t)|(u, r, t) \in T_1\} \cup \{(h, r, v)|(h, r, u) \in T_1\}$ and for G_2 are $T_2^{new} = \{(u, r, t)|(v, r, t) \in T_2\} \cup \{(h, r, u)|(h, r, v) \in T_2\}$. To clarify, the overlay graph is built by connecting the entities in seed entity pairs with edges, and the rest of the entities are connected with edges based on their similarity or co-occurrence in the training corpus. Entity embeddings are then learned using the adjacency matrix of the overlay graph and the training corpus.

Some earlier works proposed transition functions to map the embedding vectors from one KG to another, while others utilized additional information such as entity attributes to align the entity embeddings into a unified space.

2.2.3   Prediction Module

This module typically involves computing similarity scores between source and target entity embeddings and selecting the target entity with the highest score as the alignment.

To align entities, the most common method is to generate a ranked list of target entities for each source entity based on a specific distance measure between their embeddings. The distance measures commonly used include Euclidean distance, Manhattan distance, and cosine similarity. The top-ranked entity in the list is then considered a match for the source entity. It is worth noting that the similarity score can be converted into the distance score by subtracting it from 1 and vice versa.[1] In contrast, in GM-Align, the entity with the highest matching probability is aligned with the source entity.

Additionally, a recent method called CEA observes that there is a correlation between different entity alignment decisions, meaning that if a target entity is already matched to a source entity with high confidence, it is less likely to be matched to another source entity. To capture this correlation, CEA models it as a stable matching problem, and addresses the problem based on the distance measure, which decreases the number of mismatches and improves the accuracy of entity alignment.

---

[1] In this work, we use the distance between entity embeddings and the similarity between entity embeddings interchangeably.

2.2.4 Extra Information Module

In this subsection, we discuss the methods used in the extra information module.

One approach to improve the EA framework is through bootstrapping strategy, also known as iterative training or self-learning strategy. This approach involves iteratively labeling highly probable EA pairs as the training set for the next round, leading to the gradual enhancement of alignment results. There are several methods based on this approach, with variations in the selection of confident EA pairs. The approach ITransE identifies the most similar *nonaligned* target entity for each *nonaligned* source entity, and if the similarity score between them exceeds a certain threshold, they are regarded as a confident pair. BootEA, NAEA, and TransEdge follow a similar approach where they calculate the probability of each source entity being aligned with every target entity. They only consider pairs with probability scores above a certain threshold and use a maximum likelihood matching algorithm with a 1-to-1 mapping constraint to generate a set of confident EA pairs.

Several methods utilize multi-type literal information to improve alignment by providing a more comprehensive view. Commonly used types of information are the attributes associated with entities. Some methods, such as JAPE, GCN-Align, and HMAN, only consider the statistical characteristics of the attribute names. Other methods, such as AttrE and M-Greedy, generate attribute embeddings by encoding the characters of attribute values. AttrE uses attribute embeddings to unify entity embeddings into the same space, while M-Greedy uses them to complement the entity embeddings.

There is a growing tendency toward the use of "entity names".[2] Several methods are using "entity names" as input features to learn entity embeddings or exploit the semantic and string-level aspects of entity names as individual features. Specifically, GM-Align, RDGCN, and HGCN utilize entity names as input features to learn entity embeddings. On the other hand, CEA leverages both semantic and string-level aspects of entity names as individual features for alignment. Furthermore, KDCoE and the description-enhanced version of HMAN encode entity descriptions into vector representations and treat them as new features for alignment.

The availability of multi-type information is not always guaranteed in knowledge graph alignment. Some types of information like entity names are commonly available in most scenarios, while others like entity descriptions are often missing in many knowledge graphs. Additionally, due to the graph-based nature of knowledge graph alignment, most existing alignment datasets have limited textual information, which makes some approaches like KDCoE, M-Greedy, and AttrE less applicable.

---

[2] To obtain the names of entities, for DBpedia and YAGO, current approaches directly adopt the names in the identifiers, while for Wikidata, they use the entity identifier to retrieve the name of the corresponding Wikipedia page. Notably, these names from different KGs share the same naming convention.

2.3 Experiments and Analysis

This section presents an in-depth empirical study.[3]

2.3.1 Categorization

According to the main components, we can broadly categorize current methods into three groups: Group I, which merely utilizes the KG structure for alignment, Group II, which harnesses the iterative training strategy to improve alignment results, and Group III, which utilizes information in addition to the KG structure. We introduce and compare these three categories using the example in Chap. 1.

Group I This category of methods solely relies on the structure of the knowledge graph to align entities. Consider again the example in Chap. 1. In KG_{EN}, the entity Alfonso Cuarón is connected to the entity Mexico and three other entities, while Spain is connected to Mexico and one more entity. The same structural information can be observed in KG_{ES}. Since we already know that Mexico in KG_{EN} is aligned to Mexico in KG_{ES}, by using the KG structure, it is easy to conclude that the equivalent target entity for Spain is España, and the equivalent target entity for Alfonso Cuarón is Alfonso Cuarón.

Group II This category of approaches is known as iterative or self-learning strategies, where likely entity alignment pairs are labeled iteratively as the training set for the next round, leading to a progressive improvement in the alignment results. They can also be categorized into Group I or III, depending on whether they merely use the KG structure or not. Nevertheless, they are all characterized by the use of the bootstrapping strategy.

We still use the example in Chap. 1 to illustrate the bootstrapping mechanism. As shown in Fig. 1.1, by utilizing the KG structure, it is straightforward to identify that the source entity Spain is aligned with the target entity España, and the source entity Alfonso Cuarón is aligned with the target entity Alfonso Cuarón. The source entity Madrid does not have a clear target entity, as both Roma(ciudad) and Madrid in the target KG have the same structural information as the source entity. This is because they are both two hops away from the seed entity and have a degree of 1. To address this problem, bootstrapping-based approaches perform multiple rounds of alignment, using the confident entity pairs from the previous round as seed pairs for the next round. More specifically, they consider the entity pairs detected from the first round, i.e., (Spain, España) and (Alfonso Cuarón, Alfonso Cuarón), as the seed pairs in the following rounds. Consequently, in the second round, for the source entity Madrid, only

---

[3] The relevant materials are available at https://github.com/DexterZeng/EAE.

the target entity `Madrid` shares the same structural information with it—two hops away from the seed entity pair (`Mexico, Mexico`) and one hop away from the seed entity pair (`Spain, España`).

Group III Utilizing the KG structure for alignment when presented with graph-formatted input data sources is a natural choice; however, KGs also contain a wealth of semantic information that can be used to supplement structural data. These methods stand out by taking advantage of additional information beyond the KG structure.

As seen in Chap. 1, even with the KG structure and bootstrapping strategy, it is still difficult to identify the target entity for the source entity `Gravity(film)`, since its structural information (connected to the entity `Alfonso Cuarón` and with degree 2) is shared by two target entities `Gravity(película)` and `Roma(película)`. However, by combining the KG structure with the names in the identifiers, it is easy to differentiate between the two entities and correctly identify `Gravity(película)` as the target entity for `Gravity(film)`.

2.3.2 Experimental Settings

The datasets and metrics utilized for assessment were previously introduced in Chap. 1. In the following section, we will elaborate on the techniques and parameter configurations used for comparison.

Methods to Compare We will compare the previously mentioned methods, with the exception of KDCoE and MultiKE, due to the absence of entity descriptions in the evaluation benchmarks. Additionally, we will exclude AttrE since it is only functional in the mono-lingual context. Furthermore, we will provide the outcomes of the structure-only versions of JAPE and GCN-Align, specifically JAPE-Stru and GCN-Align(SE).

As previously stated in Chap. 1, to showcase the ability of ER methods in addressing EA, we will also compare with various name-based heuristics. These approaches are commonly used in related tasks [8, 18, 19], as they heavily depend on the resemblance between object names to identify equivalences. Concretely, we use the following:

- Lev aligns entities through the utilization of *Levenshtein distance* [14], which is a string-based measurement tool for computing the dissimilarity between two sequences.
- Embed aligns entities based on the cosine similarity between the averaged word embeddings, or name embeddings, of two entities. In accordance with [31], we utilize the pre-trained fastText embeddings [1] as word embeddings. For multilingual KG pairs, we use the MUSE word embeddings [7].

Implementation Details The experiments were performed using a personal computer equipped with an Intel Core i7-4790 CPU, an NVIDIA GeForce GTX TITAN X GPU, and 128 GB of memory. All programs were implemented in Python.

To ensure reproducibility, we employ the source codes provided by the authors and utilize the parameter settings specified in their original papers to execute the models.[4] For datasets not included in the original papers, we use the same parameter settings as those employed in the original experiments to ensure consistency.

All of the evaluated methods provide results on the DBP15K dataset in their original papers, with the exception of MTransE and ITransE. We compare our implemented results with the reported results from the original papers. If the difference between our results and the reported results falls outside of a reasonable range, which we define as ±5% of the original results, we mark the methods with an asterisk *. It is worth noting that there should not be a significant difference theoretically since we use the same source codes and parameter settings for implementation. For the SRPRS dataset, only RSNs reports results in its original paper [11]. We conduct experiments on all methods for SRPRS and present the results in Table 2.3. For the DWY100K dataset, we run all approaches and compare the performance of BootEA, MuGNN, NAEA, KECG, and TransEdge with the results provided in their original papers. We mark methods with notable differences with an asterisk *.

On each dataset, we highlight the best results within each group by denoting them in **bold**. We also mark the best Hits@1 performance among all approaches with ▲ since this metric is the most crucial and can best reflect the effectiveness of EA methods.

2.3.3 Results and Analyses on DBP15K

We then compare the performance within each category and across categories. The experiment results on the cross-lingual dataset DBP15K can be found in Table 2.2. Note that the Hits@10 and MRR results of CEA are missing in this table since it directly generates aligned entity pairs instead of returning a list of ranked entities.[5] We then compare the performance both within each category and across categories.

Group I Out of the methods that only utilize the KG structure, RSNs consistently obtains superior outcomes in Hits@1 and MRR metrics. This success can be attributed to its ability to capture long-term relational paths, which offer more structural indications for alignment. The performance of MuGNN and KECG is equivalent, which can be partly attributed to their shared goal of completing KGs

---

[4] In the interest of space, we put the detailed parameter settings in Appendix B.

[5] The Hits@10 and MRR results of CEA are also missing in Table 2.3 and Table 2.4 for the same reason.

Table 2.2 Experimental results on DBP15K

Method	DBP15K$_{\text{ZH-EN}}$			DBP15K$_{\text{JA-EN}}$			DBP15K$_{\text{FR-EN}}$		
	Hits@1	Hits@10	MRR	Hits@1	Hits@10	MRR	Hits@1	Hits@10	MRR
MTransE	20.9	51.2	0.31	25.0	57.2	0.36	24.7	57.7	0.36
JAPE-Stru	37.2	68.9	0.48	32.9	63.8	0.43	29.3	61.7	0.40
GCN-Align(SE)	39.8	72.0	0.51	40.0	72.9	0.51	38.9	74.9	0.51
RSNs	**58.0**	81.1	**0.66**	**57.4**	79.9	**0.65**	**61.2**	84.1	**0.69**
MuGNN	47.0	83.5	0.59	48.3	**85.6**	0.61	49.1	**86.7**	0.62
KECG	47.7	**83.6**	0.60	49.2	84.4	0.61	48.5	84.9	0.61
ITransE	33.2	64.5	0.43	29.0	59.5	0.39	24.5	56.8	0.35
BootEA*	61.4	84.1	0.69	57.3	82.9	0.66	58.5	84.5	0.68
NAEA*	38.5	63.5	0.47	35.3	61.3	0.44	30.8	59.6	0.40
TransEdge	**75.3**	**92.4**	**0.81**	**74.6**	**92.4**	**0.81**	**77.0**	**94.2**	**0.83**
JAPE	41.4	74.1	0.53	36.5	69.5	0.48	31.8	66.8	0.44
GCN-Align	43.4	76.2	0.55	42.7	76.2	0.54	41.1	77.2	0.53
HMAN	56.1	**85.9**	0.67	55.7	86.0	0.67	55.0	87.6	0.66
GM-Align*	59.5	77.9	0.66	63.5	83.0	0.71	79.2	93.6	0.85
RDGCN	69.7	84.2	0.75	76.3	**89.7**	**0.81**	87.3	95.0	0.90
HGCN	70.8	84.0	**0.76**	75.8	88.9	**0.81**	88.8	**95.9**	**0.91**
CEA	**78.7**▲	-	-	**86.3**▲	-	-	**97.2**▲	-	-
Embed	57.5	68.6	**0.61**	65.1	75.4	**0.69**	81.6	88.9	**0.84**
Lev	7.0	8.9	0.08	6.6	8.8	0.07	78.1	87.4	0.81

and reconciling structural disparities. While MuGNN utilizes AMIE+ [10] to induce rules for completion, KECG harnesses TransE to implicitly achieve this aim.

The remaining three techniques achieve comparatively lower outcomes. MTransE and JAPE-Stru leverage TransE to capture the KG structure, but JAPE-Stru outperforms MTransE because the latter models KG structures in different vector spaces, resulting in information loss when translating between them [21]. On the other hand, GCN-Align(SE) attains relatively superior results than MTransE and JAPE-Stru.

Group II Among these methods, ITransE obtains notably poorer outcomes, which can be attributed to the information loss during embedding space translation and its simpler bootstrapping strategy as described in Sect. 2.2.4. BootEA, NAEA, and TransEdge all utilize the same bootstrapping strategy. BootEA achieves slightly inferior performance compared to reported outcomes, while NAEA performs significantly worse. In theory, NAEA should outperform BootEA as it employs an attention mechanism to capture neighbor-level information. On the other hand, TransEdge employs an edge-centric embedding model to capture structural information, resulting in more accurate entity embeddings and hence better alignment outcomes.

Group III Both JAPE and GCN-Align utilize attributes to enhance entity embeddings, and their outcomes surpass those of their structure-only counterparts, demonstrating the utility of attribute information. Additionally, HMAN, which incorporates relation types as input, outperforms JAPE and GCN-Align by also utilizing attributes.

The remaining four methods utilize entity names instead of attributes for alignment and achieve superior outcomes. Among them, RDGCN and HGCN attain similar results, surpassing GM-Align. This can be attributed to their use of relations to optimize entity embedding learning, which was mostly overlooked in prior GNN-based EA models. However, CEA achieves the best performance in this group by effectively utilizing and merging available features.

Name-Based Heuristics Regarding KG pairs with closely related languages, Lev achieves encouraging results, but it is ineffective on distantly related language pairs such as DBP15K$_{ZH-EN}$ and DBP15K$_{JA-EN}$. On the other hand, Embed attains consistent performance on all KG pairs.

Intra-Category Comparison Across all datasets, CEA obtains the best Hits@1 performance, while TransEdge, RDGCN, and HGCN achieve the top results for other metrics. This confirms the effectiveness of incorporating additional information such as the bootstrapping strategy and textual information.

The performance of name-based heuristics, such as Embed, is highly competitive, surpassing most methods that do not utilize entity name information in terms of Hits@1. This indicates that conventional ER solutions can still be effective for the EA task. However, Embed still lags behind most EA methods that integrate entity name information, such as RDGCN, HGCN, and CEA.

We can also observe that methods from the first two groups, such as TransEdge, achieve consistent results across all three KG pairs. In contrast, methods that utilize entity name information, such as HGCN, achieve much better results on KG pairs with closely related languages (DBP15K$_{FR-EN}$) than those with distantly related languages (DBP15K$_{ZH-EN}$). This indicates that language barriers can hinder the use of textual information, which can, in turn, undermine the overall effectiveness of the method.

2.3.4   Results and Analyses on SRPRS

The results on SRPRS are presented in Table 2.3. Similar observations can be made as in the case of DBP15K, which we will not elaborate on. However, we can focus on the differences from DBP15K as well as the patterns specific to this dataset.

Group I The results show that the performance of the methods on the relatively sparse KGs in SRPRS is lower compared to DBP15K. However, RSNs outperforms the other methods, closely followed by KECG. It is important to note that while MuGNN achieves decent results on DBP15K, it performs much worse on SRPRS because there are no aligned relations on SRPRS, which results in the failure of rule transferring. Additionally, the sparser KG structure leads to a smaller number of detected rules.

Group II Among these solutions, TransEdge still yields consistently superior results.

Group III In contrast to GCN-Align(SE) and JAPE-Stru, incorporating attributes into GCN-Align leads to better results, but it does not contribute to the performance of JAPE. This is likely because the dataset has a relatively smaller number of attributes. On the other hand, using entity names significantly improves the results. It is worth noting that CEA achieves ground-truth performance on SRPRS$_{DBP-WD}$ and SRPRS$_{DBP-YG}$.

Name-Based Heuristics For mono-lingual EA datasets like DBpedia, Wikidata, and YAGO, Lev and Embed are able to achieve ground-truth performance since the equivalent entities in different KGs have identical names based on their entity identifiers, making it easy to achieve accurate results through a simple comparison of these names. Additionally, Lev shows promising results on cross-lingual KG pairs with closely related language pairs.

Intra-Category Comparison In contrast to DBP15K, methods that incorporate entity names (Group III) perform much better on SRPRS. This is likely due to two reasons: (1) the KG structure is less effective on this dataset, which is much sparser compared to DBP15K, and (2) the entity name information plays a significant role on both mono-lingual and cross-lingual datasets with closely related language pairs, where the names of equivalent entities are very similar.

Table 2.3 Experimental results on SRPRS

Method	SRPRS_EN-FR			SRPRS_EN-DE			SRPRS_DBP-WD			SRPRS_DBP-YG		
	Hits@1	Hits@10	MRR	Hits@1	Hits@10	MRR	Hits@1	Hits@10	MRR	Hits@1	Hits@10	MRR
MTransE	21.3	44.7	0.29	10.7	24.8	0.16	18.8	38.2	0.26	19.6	40.1	0.27
JAPE-Stru	24.1	53.3	0.34	30.2	57.8	0.40	21.0	48.5	0.30	21.5	51.6	0.32
GCN-Align(SE)	24.3	52.2	0.34	38.5	60.0	0.46	29.1	55.6	0.38	31.9	58.6	0.41
RSNs	**35.0**	**63.6**	**0.44**	**48.4**	**72.9**	**0.57**	**39.1**	**66.3**	**0.48**	**39.3**	**66.5**	**0.49**
MuGNN	13.1	34.2	0.20	24.5	43.1	0.31	15.1	36.6	0.22	17.5	38.1	0.24
KECG	29.8	61.6	0.40	44.4	70.7	0.54	32.3	64.6	0.43	35.0	65.1	0.45
ITransE	12.4	30.1	0.18	13.5	31.6	0.20	10.1	26.2	0.16	10.3	26.0	0.16
BootEA	36.5	64.9	0.46	50.3	73.2	0.58	38.4	66.7	0.48	38.1	65.1	0.47
NAEA	17.7	41.6	0.26	30.7	53.5	0.39	18.2	42.9	0.26	19.5	45.1	0.28
TransEdge	**40.0**	**67.5**	**0.49**	**55.6**	**75.3**	**0.63**	**46.1**	**73.8**	**0.56**	**44.3**	**69.9**	**0.53**
JAPE	24.1	54.4	0.34	26.8	54.7	0.36	21.2	50.2	0.31	19.3	50.0	0.30
GCN-Align	29.6	59.2	0.40	42.8	66.2	0.51	32.7	61.1	0.42	34.7	64.0	0.45
HMAN	40.0	70.5	0.50	52.8	77.8	0.62	43.3	74.4	0.54	46.1	76.5	0.56
GM-Align	57.4	64.6	0.60	68.1	74.8	0.71	76.2	83.0	0.79	80.4	83.7	0.82
RDGCN	67.2	76.7	**0.71**	77.9	**88.6**	**0.82**	97.4	99.4	0.98	99.0	**99.7**	**0.99**
HGCN	67.0	**77.0**	**0.71**	76.3	86.3	0.80	98.9	**99.9**	**0.99**	99.1	**99.7**	**0.99**
CEA	**96.2▲**	–	–	**97.1▲**	–	–	**100.0▲**	–	–	**100.0▲**	–	–
Embed	58.1	66.8	0.61	62.6	77.6	0.68	**100.0▲**	**100.0**	**1.00**	**100.0▲**	**100.0**	**1.00**
Lev	**85.1**	**90.1**	**0.87**	**86.2**	**92.1**	**0.88**	**100.0▲**	**100.0**	**1.00**	**100.0▲**	**100.0**	**1.00**

Table 2.4 Experimental results on DWY100K and DBP-FB

Method	DWY100K$_{DBP-WD}$			DWY100K$_{DBP-YG}$			DBP-FB		
	Hits@1	Hits@10	MRR	Hits@1	Hits@10	MRR	Hits@1	Hits@10	MRR
MTransE	23.8	50.7	0.33	22.7	41.4	0.29	8.5	23.0	0.14
JAPE-Stru	27.3	52.2	0.36	21.6	45.8	0.30	4.7	16.1	0.09
GCN-Align(SE)	49.4	75.6	0.59	59.8	82.9	0.68	17.8	42.3	0.26
RSNs	49.7	77.0	0.59	61.0	85.7	0.69	**25.3**	**49.7**	**0.34**
MuGNN	60.4	**89.4**	0.70	**73.9**	**93.7**	**0.81**	21.3	51.8	0.31
KECG	**63.1**	88.8	**0.72**	71.9	90.4	0.79	23.1	50.8	0.32
ITransE	17.1	36.4	0.24	15.9	36.1	0.23	3.0	10.0	0.06
BootEA*	**69.9**	86.1	0.76	61.1	77.5	0.69	21.2	42.5	0.29
TransEdge*	68.4	**90.0**	**0.79**	**83.4**	**95.3**	**0.89**	**30.4**	**56.9**	**0.39**
JAPE	33.9	60.9	0.43	21.6	45.7	0.30	6.5	20.4	0.12
GCN-Align	51.3	77.7	0.61	59.6	83.7	0.68	17.8	42.3	0.26
HMAN	65.5	89.7	0.74	77.6	93.8	0.83	25.9	54.2	0.36
GM-Align	86.3	92.2	0.89	78.3	82.3	0.80	72.1	85.5	0.77
RDGCN	–	–	–	–	–	–	67.5	84.1	0.73
HGCN	98.4	**99.2**	**0.99**	99.2	**99.9**	**0.99**	77.9	**92.3**	**0.83**
CEA	**100.0▲**	–	–	**100.0▲**	–	–	**96.3▲**	–	–
Embed	**100.0▲**	**100.0**	**1.00**	**100.0▲**	**100.0**	**1.00**	58.3	79.4	**0.65**
Lev	**100.0▲**	**100.0**	**1.00**	**100.0▲**	**100.0**	**1.00**	57.8	78.9	0.64

2.3.5 Results and Analyses on DWY100K

Table 2.4 shows the results on the large-scale mono-lingual dataset DWY100K. However, we were unable to obtain the results of RDGCN and NAEA due to their requirement for an extremely large amount of memory space in our experimental environment.

The methods in the first group perform significantly better on this dataset, which can be attributed to the relatively richer KG structure (as shown in Fig. 1.2 in Chap. 1). Among them, MuGNN and KECG achieve over 60% Hits@1 on DWY100K$_{DBP-WD}$ and over 70% on DWY100K$_{DBP-YG}$, due to the rich structure that facilitates the process of KG completion, ultimately leading to improved EA performance.

The approaches in the second group achieve further improvement in results with the aid of the iterative training strategy. However, the reported results of BootEA and TransEdge are slightly higher than the values we obtained. Among the methods in Group III, CEA achieves ground-truth performance. Similar to SRPRS, the name-based heuristics Lev and Embed also achieve ground-truth results.

Table 2.5   Averaged time cost on each dataset (in seconds)

Method	DBP15K	SRPRS	DWY100K	DBP-FB
MTransE	6,467	3,355	70,085	9,147
JAPE-Stru	298	405	6,636	767
GCN-Align(SE)	**49**	**42**	**1,446**	**103**
RSNs	7,539	2,602	28,516	7,172
MuGNN	3,156	2,215	47,735	9,049
KECG	3,724	1,800	125,386	51,280
ITransE	**494**	**175**	**9,021**	**517**
BootEA	4,661	2,659	64,471	4,345
NAEA	19,115	11,746	–	–
TransEdge	3,629	1,210	20,839	3,711
JAPE	5,578	586	21,129	1,201
GCN-Align	**103**	**87**	**3,212**	**227**
HMAN	5,455	4,424	31,895	8,878
GM-Align	26,328	13,032	459,715	53,332
RDGCN	6,711	886	–	3,627
HGCN	11,275	2,504	60,005	4,983
CEA	128	101	17,412	345

2.3.6   Efficiency Analysis

In order to provide a comprehensive evaluation, we report the average running time of each method on each dataset in Table 2.5, which allows us to compare the efficiency of different state-of-the-art solutions and provides insights into their *scalability*. We acknowledge that different parameter settings, such as the learning rate and number of epochs, may influence the final time cost. However, we aim to provide a general understanding of the efficiency of these methods by adopting the parameters reported in their original papers. As previously mentioned, we were unable to obtain the results of RDGCN and NAEA on DWY100K due to their requirement for an extremely large amount of memory space in our experimental environment.

On DBP15K and SRPRS, GCN-Align(SE) is the most efficient method with consistent alignment performance, followed closely by JAPE-Stru and ITransE. Most of the other methods have similar time costs (ranging from 1,000 to 10,000 seconds), except for NAEA and GM-Align, which require significantly longer running times.

The larger size of the DWY100K dataset leads to a significant increase in the time costs of all methods. MuGNN, KECG, and HMAN cannot run on GPUs due to memory limitations, and the authors of the original papers suggest running them on CPUs, which results in longer running times. Only three methods can complete the alignment process within 10,000s, while most of the other approaches take between 10,000s and 100,000s. In particular, GM-Align requires 5 days to generate the results, indicating that current state-of-the-art EA methods still have low efficiency when dealing with very large-scale data. Some methods, such as NAEA, RDGCN, and GM-Align, have poor scalability.

2.3.7 Comparison with Unsupervised Approaches

There exist some unsupervised methods aimed at aligning KGs that do not employ representation learning methodologies. To ensure the study's comprehensiveness, we compare with a typical system, namely, PARIS [20]. PARIS relies on the comparison of similarities between literals and employs a probabilistic algorithm to align entities jointly in an unsupervised manner. Additionally, we also evaluate PARIS alongside AgreementMakerLight (AML) [9], an unsupervised system for ontology alignment that leverages KGs' background knowledge.[6]

The F1 score is employed as the evaluation metric since PARIS and AML do not produce a target entity for every source entity, thereby addressing cases where certain entities do not have a corresponding match in the other KG. The F1 score is calculated as the harmonic mean between precision (i.e., the number of correctly aligned entity pairs divided by the number of source entities for which an approach returns a target entity) and recall (i.e., the number of source entities for which an approach returns a target entity divided by the total number of source entities).

Figure 2.2 illustrates that the overall performance of PARIS and AML is marginally lower than that of CEA. Despite CEA exhibiting more robust performance, it depends on training data (seed entity pairs) that may not be present in actual KGs. In contrast, unsupervised systems do not necessitate any training data and can still produce highly favorable outcomes. Furthermore, the results from PARIS and AML demonstrate that ontology information does, in fact, enhance the alignment outcomes.

2.3.8 Module-Level Evaluation

To obtain a better understanding of the techniques employed in various modules, we conduct an evaluation at the module level and present the associated experimental outcomes. More specifically, we select the representative methods from each module and create feasible combinations. By comparing the performance of different combinations, we can obtain a more precise assessment of the efficacy of various methods in these modules.

---

[6] AML requires ontology information, which does not exist in current EA datasets. Therefore, we mine the ontology information for these KGs. However, we can only successfully run AML on $\text{SRPRS}_{\text{EN-FR}}$ and $\text{SRPRS}_{\text{EN-DE}}$.

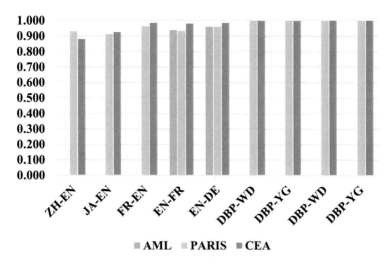

■ AML ■ PARIS ■ CEA

Fig. 2.2 F1 scores of PARIS, AML, and CEA on EA datasets

Regarding the embedding learning module, we use GCN and TransE. As for the alignment module, we adopt the margin-based loss function (Mgn) and the corpus fusion strategy (Cps). Following current approaches, we combine GCN with Mgn, and TransE with Cps, where the parameters are tuned in accordance with GCN-Align and JAPE, respectively. In the prediction module, we use the Euclidean distance (Euc), the Manhattan distance (Manh), and the cosine similarity (Cos). With regard to the extra information module, we denote the use of the bootstrapping strategy as B by implementing the iterative method in [32]. The use of multi-type information is represented as Mul, and we adopt the semantic and string-level features of entity names as in CEA.

The Hits@1 results of 24 combinations are shown in Table 2.6.[7] It is evident that the addition of the bootstrapping strategy and/or textual information does, in fact, improve the overall performance. Regarding the embedding model, the GCN+Mgn model appears to have more robust and superior performance than TransE+Cps. Furthermore, the selection of distance measures also has an impact on the outcomes. Compared with Manh and Euc, Cos leads to better performance on TransE-based models, while it brings worse results on GCN-based models. Despite this, the integration of entity name embeddings results in consistently superior performance when using the Cos distance measure.

Significantly, GCN+Mgn+Cos+Mul+B (referred to as Comb.) attains the most exceptional performance, indicating that a basic amalgamation of techniques from existing modules can lead to highly favorable alignment outcomes.

---

[7] The results on other datasets exhibit similar trends and hence are omitted in the interest of space.

Table 2.6 Hits@1 results of module-level evaluation

Method	DBP15K$_{ZH-EN}$	DBP15K$_{JA-EN}$	DBP15K$_{FR-EN}$	SRPRS$_{DBP-YG}$
GCN+Mgn+Manh	39.9	39.8	37.8	32.7
GCN+Mgn+Manh+Mul	62.9	70.3	86.6	96.5
GCN+Mgn+Manh+B	47.0	47.4	45.7	37.0
GCN+Mgn+Manh+Mul+B	64.2	71.9	87.8	97.8
GCN+Mgn+Euc	40.0	39.5	36.9	32.6
GCN+Mgn+Euc+Mul	62.9	70.8	88.3	99.9
GCN+Mgn+Euc+B	46.2	47.2	44.5	37.1
GCN+Mgn+Euc+Mul+B	64.3	72.5	89.1	100.0
GCN+Mgn+Cos	37.8	38.8	35.7	29.6
GCN+Mgn+Cos+Mul	72.1	78.6	92.9	99.9
GCN+Mgn+Cos+B	44.2	47.2	44.7	33.6
GCN+Mgn+Cos+Mul+B	**74.6**	**81.3**	**93.5**	**100.0**
TransE+Cps+Manh	41.0	36.0	30.3	18.1
TransE+Cps+Manh+Mul	65.0	72.1	86.6	94.9
TransE+Cps+Manh+B	46.4	40.9	35.7	18.8
TransE+Cps+Manh+Mul+B	66.3	73.3	87.7	95.1
TransE+Cps+Euc	41.5	36.4	30.6	18.0
TransE+Cps+Euc+Mul	64.8	71.8	87.9	99.9
TransE+Cps+Euc+B	46.3	41.2	35.7	18.2
TransE+Cps+Euc+Mul+B	65.7	72.7	88.6	100.0
TransE+Cps+Cos	41.5	36.4	30.6	18.2
TransE+Cps+Cos+Mul	71.6	77.4	92.1	99.9
TransE+Cps+Cos+B	46.5	41.4	36.0	19.1
TransE+Cps+Cos+Mul+B	73.4	78.4	92.9	100.0

2.3.9   Summary

We summarize the major findings from the experimental results.

EA vs. ER EA is distinctive from other related tasks since it operates on *graph-structured* data. As a result, all current EA solutions utilize the KG structure to create entity embeddings for aligning entities, which can produce favorable results on DBP15K and DWY100K. Nonetheless, depending solely on the KG structure has certain limitations, as there are long-tail entities with minimal structural information or entities that have similar neighboring entities but do not refer to the same real-world object. To address this issue, recent studies propose incorporating textual information, leading to better performance. However, this prompts a question regarding whether ER approaches can handle the EA task, given that the texts linked to entities are often used by conventional ER solutions.

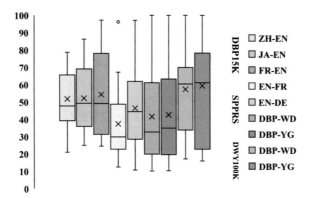

Fig. 2.3 The box plot of Hits@1 of all methods on different datasets

We answer this question by involving the name-based heuristics that have been used in most typical ER methods for comparison, and the experimental results reveal that: (1) ER solutions can indeed function on EA, but their performance is heavily reliant on the textual similarity between entities (2) While ER solutions can surpass the majority of structure-based EA methods, *they are still surpassed by EA techniques* that use name information to supplement entity embeddings (3) Incorporating the primary concepts in ER, specifically utilizing literal similarity to identify the equivalence between entities, into EA methods, is a promising direction that is worth exploring (as demonstrated by CEA)

Influence of Datasets Figure 2.3 illustrates that the performance of EA methods varies significantly across different datasets. In general, dense datasets such as DBP15K and DWY100K tend to yield relatively better results than sparse ones. Moreover, mono-lingual KGs perform better than cross-lingual ones (DWY100K vs. DBP15K). Notably, on all mono-lingual datasets, the most performant method CEA, as well as the name-based heuristics Lev and Embed, achieves 100% accuracy. This is because these datasets are sourced from DBpedia, Wikidata, and YAGO, where equivalent entities in different KGs have identical names based on their entity identifiers, making it possible to obtain ground-truth results through a simple comparison of these names. However, these datasets do not reflect the real-life challenge of ambiguous entity names. To address this, we introduce a new mono-lingual benchmark, which will be discussed in the following section.

2.3.10 Guidelines and Suggestions

In this subsection, we provide guidelines and suggestions for potential users of EA approaches.

Guidelines for Practitioners There are several considerations that may impact the selection of EA models. We have identified four of the most prevalent factors and provide the following recommendations:

- **Input information**. If the input data only includes structured information from a knowledge graph, one may need to decide between using methods from Group I or Group II. On the other hand, if there is a lot of additional information available, one may prefer to use methods from Group III to make the most of these features and generate more trustworthy signals for alignment.
- **The scale of data**. As explained in Sect. 2.3.6, certain cutting-edge techniques may not be scalable enough. Thus, it is important to consider the scale of the data before deciding on an alignment approach. For very large datasets, it may be wise to utilize simpler yet effective models, like GCN-Align, in order to minimize computational burden.
- **The objective of alignment**. When the primary focus is on aligning entities, it may be preferable to employ models based on GNNs because they tend to be more resilient and adaptable. However, if there are other tasks involved, such as aligning relations, it might be more appropriate to use KG representation-based methods as they inherently learn both entity and relation representations. Additionally, recent research studies [23, 27] indicate that relations can aid in aligning entities.
- **The trade-off in bootstrapping**. The bootstrapping process is a useful technique that can enhance the training set gradually and lead to improved alignment results. However, it can be susceptible to the problem of error propagation, which may introduce incorrectly matched entity pairs and amplify their negative effects in subsequent rounds. Additionally, it can be time-consuming. Therefore, when deciding whether to utilize the bootstrapping strategy, it is important to assess the difficulty of the datasets. If the datasets are relatively straightforward, with ample literal information and dense KG structures, utilizing the bootstrapping strategy may be a more suitable option. Otherwise, one should exercise caution when using this approach.

Suggestions for Future Research We also discuss some open problems that are worthy of exploration in the future:

- **EA for long-tail entities**. In actual knowledge graphs, most entities have few connections to other entities, while only a small number of entities have many connections. Aligning these less common entities is important for achieving good overall alignment performance, but current research on entity alignment has largely ignored them. A recent study [32] addresses this issue by using additional information to help align these less common entities and reducing their number through a KG completion process integrated into iterative self-training. However, there is still a lot of potential for further improvement in this area.
- **Multi-modal EA**. Entities can be linked to information in various forms, including texts, images, and even videos. Therefore, it is necessary to explore multi-modal entity alignment, which involves aligning entities that have multiple modalities of associated information. This topic is worth further research [16].

- **EA in the open world**. Most existing EA methods [12] operate under a closed-domain assumption, meaning that every entity in the source KG has a corresponding entity in the target KG. However, in real-world scenarios, there are always entities that cannot be matched. Moreover, labeled data, which is often necessary for state-of-the-art approaches, may not be accessible. Therefore, it is important to investigate EA in open-world settings, where unmatchable entities and limited labeled data are taken into account.

2.4   New Dataset and Further Experiments

As mentioned earlier, in current mono-lingual datasets, entities that have equivalent counterparts in different knowledge graphs have the same names based on their entity identifiers, which allows for reasonably accurate results through simple name comparison (with 100% precision on SRPRS$_{\texttt{DBP-YG}}$). However, in real-life KGs, entity identifiers are often not human-readable, and instead, they are linked to one or more human-readable names. For instance, Freebase identifies the capital of France as /m/05qtj, which is linked to names like "Paris" or "The City of Light." Retrieving these names and matching entities that share the same name can still yield a precision of 100% on datasets such as DWY100K$_{\texttt{DBP-WD}}$ and SRPRS$_{\texttt{DBP-WD}}$. However, in actual knowledge graphs, different entities can have the same name, even if they have different identifiers. For instance, the Freebase entities /m/05qtj (the capital of France) and /m/0h0_x (the king of Troy) share the name "Paris," as do 20 cities in the USA. This means that using just the entity name to match entities will not work in real-life knowledge graphs. This presents a significant challenge for EA because it is not always certain that an entity with the name "Paris" in the source knowledge graph is the same as an entity with the same name in the target knowledge graph. The reason is that one might refer to the city in France, while the other might refer to the king of Troy. This is a significant complication in real-life knowledge graphs, as illustrated by the fact that in YAGO 3, about 34% of entities share a name with one or more other entities. This problem is not fully reflected in the commonly used mono-lingual datasets for EA.

A second issue with EA datasets is that they assume that for each entity in the source KG, there is exactly one corresponding entity in the target KG. This means that an EA approach can map each source entity to the most similar target entity. However, this is not a realistic scenario since KGs in real life may contain entities that are not present in other KGs. For instance, when aligning YAGO 3 and DBpedia, some entities may appear in YAGO 3 but not in DBpedia and vice versa. This problem is even more severe for KGs that draw data from various sources, such as YAGO 4 and IMDB. In YAGO 4, only 1% of entities are related to movies, while the remaining 99% are unrelated to IMDB entities, such as universities and smartphone brands. As a result, these entities have no matches in IMDB, and this problem is not addressed in current EA datasets.

We thus observe that the existing datasets for EA are an oversimplification of the real-life problem. Our solution is to create a fresh dataset that mimics these challenges. We anticipate that this dataset will result in improved EA models that can handle even more demanding problem scenarios and provide a clearer research direction for the community. In this section, we describe the development of the new dataset and present our experimental findings on it.

2.4.1 Dataset Construction

To reflect the difficulty of using entity names, we choose Freebase [2] as our target knowledge graph because it represents entities using indecipherable identifiers (i.e., Freebase MIDs), and different entities may have the same name. As to the source knowledge graph, we utilize DBpedia, which contains external links to Freebase that can be regarded as gold standards. The detailed process of constructing the new dataset is explained below:

Determining the Source Entity Set We utilize the disambiguation information available in DBpedia to gather entities that have the same disambiguation term and create the entity set for the source knowledge graph. For example, for the ambiguous term *Apple*, the disambiguation records consist of entities such as `Apple Inc.` and `Apple(fruit)`, both of which are included in the source entity set.

Determining Links and the Target Entity Set Next, we utilize the external links between DBpedia and Freebase to obtain the entities in Freebase that correspond to the source entities and create the entity set for the target knowledge graph. These external links are considered as the gold standards. It should be noted that the entities in the target knowledge graph are identified using Freebase MIDs and multiple entities may have the same name, such as `Apple`. To retrieve the name for each entity, we use the label triples.

Retrieving Triples Once the entity sets for the source and target knowledge graphs are determined, we extract the relational and attributive triples involving these entities from their respective knowledge graphs.

Refining Links and Entity Sets Following the approach in previous work [21, 22], we retain only the links whose source and target entities are involved in at least one triple in their respective knowledge graphs, resulting in a total of 25,542 links. The entity sets are adjusted accordingly, including entities that participate in triples but not in links. Ultimately, there are 29,861 entities in the source knowledge graph, of which 4,319 cannot be matched, and 25,542 matchable entities in the target knowledge graph. Consistent with existing datasets, 30% of the links and unmatchable entities are utilized as the training set. For additional statistics on the dataset, please refer to Chap. 1.

2.4.2 Experimental Results on DBP-FB

In accordance with the current evaluation paradigm, we first analyze the performance of EA methods without considering unmatchable entities. As shown in Table 2.4, the overall performance of the methods in the first two groups is lower than that on SRPRS. This can be attributed to the greater structural heterogeneity of DBP-FB, which can be observed from sub-figures (d) in Fig. 1.2. In contrast to the KG pairs in sub-figures (a), (b), or (c), the entity distributions in these KGs are highly dissimilar, which makes it challenging to effectively leverage the structural information.

Methods that utilize entity names continue to produce the best results, although their performance is lower than that on previous mono-lingual datasets. Furthermore, on DBP-FB, Embed and Lev achieve only Hits@1 values of 58.3% and 57.8%, respectively, while they attain 100% on $SRPRS_{DBP-YG}$, $SRPRS_{DBP-WD}$, $DWY100K_{DBP-YG}$, and $DWY100K_{DBP-WD}$. This confirms that DBP-FB is a more suitable mono-lingual dataset for addressing the challenge of entity name ambiguity compared to existing datasets. Thus, DBP-FB can be considered a preferable mono-lingual dataset.

2.4.3 Unmatchable Entities

In addition, DBP-FB also contains unmatchable entities, which presents another real-life challenge for EA. We therefore evaluate the performance of Comb. (from Sect. 2.3.8) on DBP-FB, taking into account these unmatchable entities. Consistent with Sect. 2.3.7, we utilize the *precision*, *recall*, and *F1* score as evaluation metrics, with the exception that we define *recall* as the number of matchable source entities for which an approach returns a target entity, divided by the total number of matchable source entities.

The information presented in Table 2.7 shows that Comb. exhibits a high level of recall, but its precision is relatively low. This is because it creates a target entity for each source entity, including those that cannot be matched. This pattern reflects the current performance of entity alignment solutions when dealing with unmatchable source entities. Nonetheless, this problem is not addressed in the current entity alignment datasets.

In order to address this issue, we suggest a straightforward approach to handle unmatchable entities in DBP-FB, in addition to the current entity alignment solutions. Specifically, we propose setting a NIL threshold, denoted as θ, to predict unmatchable entities. As discussed in Sect. 2.2.3, entity alignment solutions typically employ a distance measure to find the corresponding target entity. If the distance value between a source entity and its nearest target entity is greater than θ, we consider the source entity to be unmatchable and exclude it from the alignment results. The value of the threshold θ can be determined from the training data.

Table 2.7 EA performance
on DBP-FB after considering
unmatchable entities

Method	Precision	Recall	F1
Comb.	0.667	**1.000**	0.800
Comb. +TH	**0.790**	0.908	**0.845**

As shown in Table 2.7, the threshold-enhanced solution Comb. +TH achieves a better F1 score. We hope this preliminary study can inspire follow-up research on this issue.

2.5 Conclusion

Entity alignment plays a crucial role in integrating KGs to enhance knowledge coverage and quality. Despite the numerous proposed solutions, there has been limited comprehensive evaluation and detailed analysis of their performance. To address this gap, this chapter presents an empirical assessment of state-of-the-art approaches in terms of effectiveness and efficiency on representative datasets. We also conduct a thorough analysis of their performance and provide evidence-based discussions. Furthermore, we introduce a new dataset that more accurately reflects real-world challenges, which can serve as a benchmark for future research in this field.

Appendix A

Methods in Group I of Table 2.1

MTransE The MTransE model [6] is a translation-based approach for learning multilingual KG embeddings to support EA. Initially, it utilizes TransE (without negative triples) to project each KG into separate embedding spaces. Next, MTransE applies three distinct transition strategies: distance-based axis calibration, translation vectors, and linear transformation, to map the embedding vectors to their cross-lingual counterparts. During the prediction stage, a KNN search is conducted on the cross-lingual transition point of a target entity to obtain its corresponding counterpart.

RSNs In this study [11], RNNs are combined with residual learning to effectively capture the long-term relational dependencies among KGs and generate more comprehensive KG structural embeddings for EA.

The paper [11] argues that triple-level learning is inadequate for capturing the long-term relational dependencies of entities and for propagating semantic information among entities. To address this limitation, recurrent skipping networks

(RSNs) are proposed to learn the long-term relational paths between entities. To obtain the desired paths, biased random walks are used to efficiently sample paths from the KGs, with elements in two KGs connected by seed alignments. During the prediction phase, cosine similarity is utilized to predict the results.

MuGNN The paper [4] proposes a multichannel GNN for learning KG embeddings that are oriented toward entity alignment.

The MuGNN approach first conducts relation weighting to generate a weight matrix for each KG using KG self-attention and cross-KG attention schemes, which correspond to different GNN channels. Next, it applies the GNN encoder (and the corresponding weight matrix) for each channel to model the KG structure. The outputs of the different channels are combined using the pooling operation. Finally, a margin-based alignment model is utilized to embed the two KGs into a unified embedding space.

In addition, MuGNN also proposes a method to address structural differences between KGs by completing missing relations. This is accomplished by using AMIE+ to induce rules and transferring rules between KGs through aligned relations. However, it is important to note that not all datasets have aligned relations, which may cause the rule transferring approach to fail.

KECG The paper [15] proposes a method for jointly learning a knowledge embedding model that encodes inner-graph relationships and a cross-graph model that enhances entity embeddings with their neighbors' information.

The main concept behind KECG involves employing a cross-graph model, which is an enhanced version of a graph attention network (GAT), to convert entities into a single vector space by incorporating both intra-graph and inter-graph alignment information. The resulting embeddings are then utilized as input for TransE, which models intra-graph connections and enforces relational constraints between entities to promote consistency across different KGs. During inference, equivalent entities are identified based on the L2 distance between entities in terms of the unified embeddings.

Methods in Group II of Table 2.1

ITransE This study [34] extends the use of TransE to learn the structure of knowledge graphs. It develops three models, including translation-based, linear transformation-based, and parameter sharing-based models, to generate joint embeddings for various knowledge graphs. The study proceeds to iteratively align entities and update the joint knowledge embeddings, progressively considering highly confident aligned entities identified by the model. During the prediction stage, the model retrieves the closest entity from the target knowledge graph as the corresponding entity for each source entity.

BootEA This work [22] suggests a technique called bootstrapping for EA that involves the iterative labeling of probable EA pairs as training data to teach alignment-oriented KG embeddings.

In terms of the KG structure encoder, BootEA employs TransE, but substitutes the margin-based loss function with a limit-based objective function. The approach involves learning alignment-oriented KG embeddings by swapping aligned entities between triples from different KGs. Additionally, the authors develop a bootstrapping strategy to refine alignment-oriented embeddings, which involves iteratively labeling probable alignments and adding them to the training data. BootEA further models EA as a classification problem and aims to maximize alignment likelihood across all labeled and unlabeled entities based on KG embeddings. During the prediction stage, cosine similarity is used to identify latent aligned entities.

NAEA This paper [35] introduces a technique called neighborhood-aware attentional representation to enhance the effectiveness of EA, which is built on the fundamental framework of BootEA.

NAEA comprises two components: a knowledge embedding (KE) component and an entity alignment (EA) component. KE employs an attention mechanism to obtain neighbor-level representations of entities by combining their neighbors with weighted attention, and subsequently utilizes TransE to model both neighbor-level and relation-level representations. In contrast, BootEA only encodes relation-level representations.

Like BootEA, NAEA also treats the alignment task as a classification problem in its EA component. However, in NAEA, alignment probability calculation also incorporates neighbor-level knowledge information. During the prediction stage, the approach employs cosine similarity to identify aligned entity pairs based on integrated representations of entities.

TransEdge This work [23] introduces a new edge-centric embedding model for EA, which contextualizes relation representations with respect to particular head-tail entity pairs.

The proposed method, TransEdge, defines a novel energy function to evaluate the accuracy of edge translation between entity embeddings for KG embedding learning. To model edge embeddings, two methods are employed: context compression and context projection. The limit-based loss function of TransEdge is used to optimize entity embeddings for EA, and the distance between seed entities is minimized to reconcile two KGs. During the prediction phase, the model ranks entities in another KG based on the cosine similarity of their entity embeddings in descending order for a given entity to be aligned. The intended match is expected to have the highest rank.

Methods in Group III of Table 2.1

JAPE This work [21] presents a joint attribute-preserving embedding model for EA, which generates embeddings that incorporate both KG relations and attributes.

The proposed JAPE approach first employs TransE to encode the structure of each KG, but adapts the loss function. In addition to the large margin between scores of positive and negative triples, JAPE aims to assign lower scores to positive triples and higher scores to negative triples. The seed EA pairs are used to construct an overlay relationship graph in the corpus, which can align separate KG embeddings into a unified one.

Additionally, JAPE observes that latent aligned entities tend to have similar attribute values, and therefore abstracts attribute values to their range types and generate attribute embeddings to capture attribute correlations. Finally, attribute similarity constraints are combined with structural embeddings to refine entity representations by clustering entities with high attribute correlations. During the search for latent aligned entities, the model uses cosine similarity between entity embeddings.

GCN-Align This work [26] utilizes GCN as the KG structure encoder for aligning entities.

To elaborate further, GCN-Align leverages GCN to capture the structure information of KGs, which generates neighborhood-aware embeddings of entities. Additionally, it embeds the attribute names of entities to provide a complementary view. The model uses a margin-based ranking loss function to unify embeddings from different KGs. The structural and attributive embeddings are then combined to predict aligned entity pairs based on the Manhattan distance score. Finally, the model predicts latent entity alignments based on the distance measure between entities from the two KGs.

AttrE This work [24] proposes to learn attribute embeddings of entities, which shift the entity embeddings of two KGs into the same vector space.

First, AttrE creates a module for matching the predicates of two KGs, renaming them into a shared naming system to make sure the relation embeddings are compatible. Subsequently, TransE is used to learn the structural embeddings and attributes are encoded as attribute character embeddings. Transitivity rule is used to enrich the attribute triples. Finally, the attribute character embeddings are used to project the structural embeddings of entities into the same vector space and cosine similarity is used to make the prediction.

KDCoE This work [5] develops a semi-supervised cross-lingual method to align multilingual KGs with minimal supervision.

The KDCoE approach uses TransE as the structure encoder and combines it with a linear transformation-based network to bring together different knowledge graph embeddings. Additionally, it uses an attentive gated recurrent unit encoder (AGRU) to create representations of entity descriptions. The KDCoE approach trains both modules simultaneously, with both models suggesting a set of the most confident entity alignment pairs during each iteration to improve cross-lingual learning accuracy over time. Similarly to MTransE, the prediction is made through a KNN search from the cross-lingual conversion point of a target entity.

HMAN The HMAN [30] method utilizes GCN to merge multiple types of information in order to generate entity embeddings. Additionally, it proposes a modified model that incorporates the textual descriptions of entities, which are encoded using a pre-trained multilingual BERT model.

In detail, the HMAN approach employs GCN to model the structural connections and uses feedforward neural networks to generate embeddings for attributes and relations, as using GCN to learn attribute and relation embeddings inherently considers the neighboring entities' attributes and relations, which could lead to noise. The approach then concatenates these representations to form a hybrid multi-aspect entity embedding. Finally, the method utilizes a margin-based ranking loss function to align the entities.

Furthermore, the HMAN method introduces two additional techniques, pointwise-BERT and pairwise-BERT, for utilizing multilingual BERT on entity descriptions to aid in the entity alignment process. To integrate entity descriptions with the hybrid multi-aspect entity embeddings, two strategies, reranking and weighted concatenation, are proposed. For prediction, the method leverages the L1 distance between entity embeddings.

GM-Align The work described in reference [29] approaches the entity alignment problem as a graph matching challenge that can be addressed through both entity-level and graph-level matching techniques.

The GM-Align method initially generates a topic entity graph to depict the connections between a given entity (the "topic entity") and its neighboring entities. This graph is then used to apply GCN to encode the structural information and generate matching scores. The method employs a word-based LSTM to embed the entity names as an initial feature matrix for GCN. The matching framework learns alignment information between the two knowledge graphs. During prediction, the method ranks all entities in the other knowledge graph in descending order of their matching probabilities, with the top-ranked entity considered as the result.

RDGCN The authors of reference [27] propose a relation-aware dual-graph convolutional network to include relation information by employing attentive interactions between a knowledge graph and its dual relation counterpart, so as to achieve an effective entity alignment process.

The RDGCN method acknowledges that GCN-based models often disregard the relation information present in knowledge graphs. To address this, the authors employ the dual-primal graph CNN (DPGCNN) method to incorporate relation information. To adapt DPGCNN to the entity alignment task, the RDGCN method proposes a weighted model and explores the head/tail representations, which are initialized with entity names, as a way to capture the relation information.

The RDGCN method permits multiple rounds of interactions between the primal entity graph and its dual relation graph, thus allowing the model to integrate more complex relation information into entity representations effectively. The method employs GCN with highway gates to incorporate neighboring structural information. The authors devise a margin-based scoring function to align embeddings from different knowledge graphs. During prediction, the method uses the Manhattan distance between entity embeddings.

HGCN The authors of reference [28] suggest jointly learning entity and relation representations for the entity alignment task.

The HGCN method first uses highway-GCNs that employ highway gates to control noise propagation in GCN to embed entities from various knowledge graphs. Next, the entity embeddings are utilized to approximate relation representations, which are then used to align relations across knowledge graphs. Finally, HGCN incorporates the relation representations into the entity embeddings to obtain joint entity representations and continues to use GCN to iteratively integrate neighboring structural information to improve the entity and relation representations further. Similar to RDGCN, a margin-based scoring function is used to align embeddings from different knowledge graphs, and the entity name is used as the initial feature matrix for GCN. During prediction, the method employs the Manhattan distance between entity embeddings.

MultiKE The MultiKE method proposes a new framework that integrates entity names, relations, and attributes to learn embeddings for alignment [33].

The MultiKE method defines three different perspectives for EA, namely, entity name, relation, and attribute, and employs specific models to learn embeddings for each perspective. The TransE model is used to encode KG structure, with logistic loss replacing the margin-based loss. Two cross-KG identity inference strategies are proposed to capture and propagate alignment information between KGs. The view-specific entity embeddings are then combined, which are used for prediction through nearest-neighbor search. It should be noted that this method is currently only applicable to mono-lingual EA.

CEA The authors of [31] create a unified EA framework that takes into account how different EA decisions are interconnected.

CEA uses three types of features (structural, semantic, and string signals) to capture different aspects of entities in heterogeneous knowledge graphs. The authors then model the problem of making collective EA decisions by framing it as a stable matching problem, which is solved using the deferred acceptance algorithm.

Appendix B

Parameter Setting

The definitions of the parameters can be found in their original papers.

- MTransE: $\lambda = 0.01$, $\alpha = 5$, $k = 75$, and $epoch = 1000$.
- JAPE-Stru, JAPE: $d = 75$, $\alpha = 0.1$, $\beta = 0.05$, and $\delta = 0.05$. For SE, learning rate is set to 0.01 and early stopping. For AE, learning rate is set to 0.1 and epochs are set to 100.
- GCN-Align(SE), GCN-Align: $d_s = 300$, $d_a = 100$, $\gamma_s = 3$, $\gamma_a = 3$, and $\beta = 0.9$.
- RSNs: $\alpha = 0.9$ and $\beta = 0.9$; learning rate is set to 0.003, embedding size set to 256, batch size set to 512, and length set to 15.
- MuGNN: $\gamma_1 = 1.0$, $\gamma_2 = 1.0$, and $\gamma_r = 0.12$; embedding size is set to 128, learning rate set to 0.001, L2 set to 0.01, dropout set to 0.2, and epoch set to 500.
- KECG: $K_1 = 25$, $K_2 = 2$, $\lambda = 0.005$, $\gamma_1 = 3.0$, $\gamma_2 = 3.0$, dimension set to 128, and epoch set to 1,000.
- ITransE: $n = 50$, $\gamma = 1.0$, $k = 1.0$, $\lambda = 0.001$, and $epoch = 3000$.
- BootEA: $\gamma_1 = 0.01$, $\gamma_2 = 2.0$, $\gamma_3 = 0.7$, $\mu_1 = 0.2$, $\mu_2 = 0.1$. For DBP15K and SRPRS, $\epsilon = 0.9$; for DWY100K, $\epsilon = 0.98$. Learning rate is set to 0.01, epoch set to 50, and dimension set to 75.
- NAEA: $m = 75$, $\beta = 0.8$, $\lambda = 1$, $\mu_1 = 1$, $\mu_2 = 0.1$, $\gamma = 2$, $K = 4$, $\eta = 0.01$, and $epoch = 50$.
- TransEdge: $\gamma_1 = 0.2$, $\gamma_2 = 2.0$, $\alpha = 0.8$, $s = 0.7$, $d = 75$; learning rate is set to 0.01 and early stopping.
- HMAN: $F = 1,000$, $\beta = 3$, $\tau = 0.8$, $epoch = 50,000$. For DBP15K and SRPRS, topological, relation, and attribute embeddings are set to 200, 100, and 100, respectively; for DWY100K, dimensions are set to 100, 50, and 50, respectively.
- GM-Align: $K_1 = 2$, and $K_2 = 3$; learning rate is set to 0.001 and batch size set to 32.
- RDGCN: $\beta_1 = 0.1$, $\beta_2 = 0.3$, $\gamma = 1.0$, $d = 300$, $d' = 600$, $\tilde{d} = 300$, and $\kappa = 125$; learning rate is set to 0.001.
- HGCN: $\gamma = 1$, $\beta = 20$, and $\kappa = 125$; learning rate is set to 0.001.

References

1. P. Bojanowski, E. Grave, A. Joulin, and T. Mikolov. Enriching word vectors with subword information. *Transactions of the Association for Computational Linguistics*, 5:135–146, 2017.
2. K. D. Bollacker, C. Evans, P. Paritosh, T. Sturge, and J. Taylor. Freebase: a collaboratively created graph database for structuring human knowledge. In *SIGMOD*, pages 1247–1250, 2008.

3. A. Bordes, N. Usunier, A. García-Durán, J. Weston, and O. Yakhnenko. Translating embeddings for modeling multi-relational data. In *NIPS*, pages 2787–2795, 2013.
4. Y. Cao, Z. Liu, C. Li, Z. Liu, J. Li, and T. Chua. Multi-channel graph neural network for entity alignment. In *ACL*, pages 1452–1461, 2019.
5. M. Chen, Y. Tian, K. Chang, S. Skiena, and C. Zaniolo. Co-training embeddings of knowledge graphs and entity descriptions for cross-lingual entity alignment. In *IJCAI*, pages 3998–4004, 2018.
6. M. Chen, Y. Tian, M. Yang, and C. Zaniolo. Multilingual knowledge graph embeddings for cross-lingual knowledge alignment. In *IJCAI*, pages 1511–1517, 2017.
7. A. Conneau, G. Lample, M. Ranzato, L. Denoyer, and H. Jégou. Word translation without parallel data. *arXiv preprint arXiv:1710.04087*, 2017.
8. S. Das, P. S. G. C., A. Doan, J. F. Naughton, G. Krishnan, R. Deep, E. Arcaute, V. Raghavendra, and Y. Park. Falcon: Scaling up hands-off crowdsourced entity matching to build cloud services. In *SIGMOD*, pages 1431–1446, 2017.
9. D. Faria, C. Pesquita, E. Santos, I. F. Cruz, and F. M. Couto. Agreementmakerlight 2.0: Towards efficient large-scale ontology matching. In M. Horridge, M. Rospocher, and J. van Ossenbruggen, editors, *ISWC*, volume 1272 of *CEUR Workshop Proceedings*, pages 457–460. CEUR-WS.org, 2014.
10. L. Galárraga, C. Teflioudi, K. Hose, and F. M. Suchanek. Fast rule mining in ontological knowledge bases with AMIE+. *VLDB J.*, 24(6):707–730, 2015.
11. L. Guo, Z. Sun, and W. Hu. Learning to exploit long-term relational dependencies in knowledge graphs. In *ICML*, pages 2505–2514, 2019.
12. S. Hertling and H. Paulheim. The knowledge graph track at OAEI - gold standards, baselines, and the golden hammer bias. In A. Harth, S. Kirrane, A. N. Ngomo, H. Paulheim, A. Rula, A. L. Gentile, P. Haase, and M. Cochez, editors, *ESWC*, volume 12123 of *Lecture Notes in Computer Science*, pages 343–359. Springer, 2020.
13. T. N. Kipf and M. Welling. Semi-supervised classification with graph convolutional networks. *CoRR*, abs/1609.02907, 2016.
14. V. I. Levenshtein. Binary codes capable of correcting deletions, insertions, and reversals. In *Soviet physics doklady*, volume 10, pages 707–710, 1966.
15. C. Li, Y. Cao, L. Hou, J. Shi, J. Li, and T.-S. Chua. Semi-supervised entity alignment via joint knowledge embedding model and cross-graph model. In *EMNLP*, pages 2723–2732, 2019.
16. Y. Liu, H. Li, A. García-Durán, M. Niepert, D. Oñoro-Rubio, and D. S. Rosenblum. MMKG: multi-modal knowledge graphs. In P. Hitzler, M. Fernández, K. Janowicz, A. Zaveri, A. J. G. Gray, V. López, A. Haller, and K. Hammar, editors, *ESWC*, volume 11503 of *Lecture Notes in Computer Science*, pages 459–474. Springer, 2019.
17. F. Monti, O. Shchur, A. Bojchevski, O. Litany, S. Günnemann, and M. M. Bronstein. Dual-primal graph convolutional networks. *CoRR*, abs/1806.00770, 2018.
18. S. Mudgal, H. Li, T. Rekatsinas, A. Doan, Y. Park, G. Krishnan, R. Deep, E. Arcaute, and V. Raghavendra. Deep learning for entity matching: A design space exploration. In *SIGMOD*, pages 19–34, 2018.
19. V. Rastogi, N. N. Dalvi, and M. N. Garofalakis. Large-scale collective entity matching. *PVLDB*, 4(4):208–218, 2011.
20. F. M. Suchanek, S. Abiteboul, and P. Senellart. PARIS: probabilistic alignment of relations, instances, and schema. *PVLDB*, 5(3):157–168, 2011.
21. Z. Sun, W. Hu, and C. Li. Cross-lingual entity alignment via joint attribute-preserving embedding. In *ISWC*, pages 628–644, 2017.
22. Z. Sun, W. Hu, Q. Zhang, and Y. Qu. Bootstrapping entity alignment with knowledge graph embedding. In *IJCAI*, pages 4396–4402, 2018.
23. Z. Sun, J. Huang, W. Hu, M. Chen, L. Guo, and Y. Qu. Transedge: Translating relation-contextualized embeddings for knowledge graphs. In *ISWC*, pages 612–629, 2019.
24. B. D. Trisedya, J. Qi, and R. Zhang. Entity alignment between knowledge graphs using attribute embeddings. In *AAAI*, pages 297–304, 2019.

25. A. Vaswani, N. Shazeer, N. Parmar, J. Uszkoreit, L. Jones, A. N. Gomez, L. Kaiser, and I. Polosukhin. Attention is all you need. In *NIPS*, pages 5998–6008, 2017.
26. Z. Wang, Q. Lv, X. Lan, and Y. Zhang. Cross-lingual knowledge graph alignment via graph convolutional networks. In *EMNLP*, pages 349–357, 2018.
27. Y. Wu, X. Liu, Y. Feng, Z. Wang, R. Yan, and D. Zhao. Relation-aware entity alignment for heterogeneous knowledge graphs. In *IJCAI*, pages 5278–5284, 2019.
28. Y. Wu, X. Liu, Y. Feng, Z. Wang, and D. Zhao. Jointly learning entity and relation representations for entity alignment. In *EMNLP*, pages 240–249, 2019.
29. K. Xu, L. Wang, M. Yu, Y. Feng, Y. Song, Z. Wang, and D. Yu. Cross-lingual knowledge graph alignment via graph matching neural network. In *ACL*, pages 3156–3161, 2019.
30. H.-W. Yang, Y. Zou, P. Shi, W. Lu, J. Lin, and S. Xu. Aligning cross-lingual entities with multi-aspect information. In *EMNLP*, pages 4422–4432, 2019.
31. W. Zeng, X. Zhao, J. Tang, and X. Lin. Collective entity alignment via adaptive features. In *ICDE*, pages 1870–1873. IEEE, 2020.
32. W. Zeng, X. Zhao, W. Wang, J. Tang, and Z. Tan. Degree-aware alignment for entities in tail. In *SIGIR*, pages 811–820. ACM, 2020.
33. Q. Zhang, Z. Sun, W. Hu, M. Chen, L. Guo, and Y. Qu. Multi-view knowledge graph embedding for entity alignment. In *IJCAI*, pages 5429–5435, 2019.
34. H. Zhu, R. Xie, Z. Liu, and M. Sun. Iterative entity alignment via joint knowledge embeddings. In *IJCAI*, pages 4258–4264, 2017.
35. Q. Zhu, X. Zhou, J. Wu, J. Tan, and L. Guo. Neighborhood-aware attentional representation for multilingual knowledge graphs. In *IJCAI*, pages 1943–1949, 2019.

Part II
Recent Advances

Chapter 3
Recent Advance of Representation Learning Stage

Abstract Over the last few years, there are a pile of research devoted to learning better KG representations to facilitate entity alignment. Thus, in this chapter, we summarize recent progress in the representation learning stage of EA and also provide a detailed empirical evaluation to reveal the strengths and weaknesses of current solutions.

3.1 Overview

To better understand current advanced representation learning methods, we propose a general framework to describe these methods, which includes six modules, i.e., pre-processing, messaging, attention, aggregation, post-processing, and loss function. In pre-processing, the initial entity and relation representations are generated. Then, KG representations are obtained via a representation learning network, which usually consists of three steps, i.e., messaging, attention, and aggregation. Among them, messaging aims to extract the features of the neighboring elements, attention aims to estimate the weight of each neighbor, and aggregation integrates the neighboring information with attention weights. Through the post-processing operation, the final representations are obtained. The whole model is then optimized by the loss function in the training stage.

More specifically, we summarize ten representative methods in terms of these modules in Table 3.1.

- In the pre-processing module, there are mainly two ways to obtain the initial representations, some methods utilize pre-trained model to embed names or descriptions into initial representations, while some methods generate the initial structural representations through GNN-based networks.
- In the messaging module, linear transformation is the most frequently used strategy, which makes use of a learnable matrix to transform neighboring features. Other methods include extracting neighboring features by concatenating multi-head messages, directly utilizing neighboring representations, etc.
- In the attention module, the main focus is the computation of similarity. Most of the methods concatenate the representations and multiply a learnable attention

X. Zhao et al., *Entity Alignment*, Big Data Management,
https://doi.org/10.1007/978-981-99-4250-3_3

Table 3.1 Overview and comparison of advanced representation learning

Model	Pre-processing	Messaging	Attention	Aggregation	Post-processing	Loss function
AliNet [12]	–	Linear transformation	Inner product	1-hop and 2-hop ent.	Concat.	TransE and Margin loss
MRAEA [8]	GNN	No transformation	Concat.	1-hop ent.	Concat.	Margin loss
RREA [9]	–	Linear transformation	Concat.	1-hop ent.	Concat.	Margin loss
RPR-RHGT [2]	Pre-trained	Concat.	Concat.	1-hop ent. and rel.	Adaptive	Margin loss
RAGA [17]	Pre-trained and GNN	Linear transformation	Concat.	1-hop ent. and rel.	Concat.	Margin loss
Dual-AMN [7]	–	Linear transformation	Concat.	1-hop ent. and Proxy vectors	Concat. and Adaptive	Improved margin loss
ERMC [14]	Pre-trained	Linear transformation	–	1-hop ent. and rel.	Concat.	Sinkhorn
KE-GCN [15]	–	Linear transformation	–	1-hop ent. and rel.	–	Margin loss
RePS [13]	–	Weighted	–	Anchor set	Adaptive	Margin loss
SDEA [16]	Pre-trained	GRU	Inner product	1-hop ent.	Concat.	Margin loss

vector to calculate attention weights. Besides, some use inner product of entity representations to compute similarity.

- In the aggregation module, almost all methods aggregate 1-hop neighboring entity or relation information, while a few works propose to combine multi-hop neighboring information. Some use a set of randomly chosen entities, i.e., anchor set, to obtain position-aware representations.
- In the post-processing module, most of the methods enhance final representations by concatenating the outputs of all layers of GNN. Besides, some methods propose to combine the features adaptively via strategies such as the gate mechanism [10].
- In terms of the loss function, the majority of methods utilize the margin-based loss during training. Some additionally add the TransE [1] loss, while some improve the margin loss using LogSumExp and normalization operation, or utilizing the Sinkhorn [3] algorithm to calculate the loss.

3.2 Models

We use Eq. (3.1) to characterize the core procedure of representation learning:

$$e_i^l = \mathbf{Aggregation}_{\forall j \in \mathcal{N}(i)}(\mathbf{Attention}(i, j) \cdot \mathbf{Messaging}(i, j)) , \quad (3.1)$$

where **Messaging** aims to extract the features of neighboring elements, **Attention** aims to estimate the weight of each neighbor, and **Aggregation** integrates the neighborhood information with attention weights.

Next, we briefly introduce recent advance of representation learning for EA in terms of the modules mentioned in Table 3.1.

3.2.1 ALiNet

It aims to aggregate multi-hop structural information for learning entity representations [12].

Aggregation This work devises a multi-hop aggregation strategy. For 2-hop aggregation, **Aggregate** is denoted as:

$$h_{i,2}^l = \sigma \left(\sum_{j \in \mathcal{N}_2 \cup i} \mathbf{Attention}(i, j) \cdot \mathbf{Messaging}(i, j) \right) , \quad (3.2)$$

where $\mathcal{N}_2$ denotes the 2-hop neighbors.

Then, it aggregates the multi-hop aggregation results to generate the entity representation. Aggregating 1-hop and 2-hop information is denoted as:

$$h_i = g\left(h^l_{i,2}\right) \cdot h^l_{i,1} + \left(1 - g(h^l_{i,2})\right) \cdot h^l_{i,2}, \tag{3.3}$$

where $g(h^l_{i,2}) = \sigma(M h^l_{i,2} + b)$, which is the gate to control the influences of different hops. M and b are learnable parameters.

Attention Regarding the attention weight, it assumes that not all distant entities contribute positively to the characterization of the target entity representation, and the softmax function is used to produce the attention weights:

$$\mathbf{Attention}(i, j) = \alpha^l_{ij} = softmax\left(c^l_{ij}\right) = \frac{\exp\left(c^l_{ij}\right)}{\sum_{n \in N_2(i) \cup i} \exp\left(c^l_{in}\right)}, \tag{3.4}$$

where $c^l_{ij} = LeakyReLU((M^l_1 h^l_i)^T M^l_2 h^l_j)$, and M_1, M_2 are two learnable matrices.

Messaging The extraction of the features of neighboring entities is implemented as a simple linear transformation: $\mathbf{Messaging}(i, j) = W^l_q h^{l-1}_j$, where W_q denotes the weight matrix for the q-hop aggregation.

Post-processing The representations of all layers are concatenated to produce the final entity representation:

$$h_i = \oplus^L_{l=1} norm\left(h^l_i\right). \tag{3.5}$$

Loss Function The loss function is formulated as:

$$\mathcal{L} = \sum_{(i,j) \in \mathcal{A}^+} ||h_i - h_j|| + \sum_{(i',j') \in \mathcal{A}^-} \alpha_1 [\gamma - ||h_{i'} - h_{j'}||]_+, \tag{3.6}$$

where $\mathcal{A}^-$ is the set of negative samples, obtained through random sampling. $|| \cdot ||$ denotes the L2 norm. $[\cdot]_+ = \max(0, \cdot)$.

3.2.2 MRAEA

It proposes to utilize the relation information to facilitate the entity representation learning process [8].

Pre-processing Specifically, it first creates an inverse relation for each relation, resulting in the extended relation set $\mathcal{R}$. Then, it generates the initial features for entities by averaging and concatenating the embeddings of neighboring entities and relations:

$$h_{e_i}^{in} = \left[\frac{1}{|\mathcal{N}_i^e| + 1} \sum_{e_j \in \mathcal{N}_i^e \cup e_i} h_{e_j} || \frac{1}{|\mathcal{N}_i^r|} \sum_{r_k \in \mathcal{N}_i^r} h_{r_k} \right], \tag{3.7}$$

where the embeddings of entities and relations are randomly initialized.

Aggregation The aggregation is a simple combination of the extracted features and the weights:

$$h_{e_i}^{out} = \sigma \left(\sum_{e_j \in \mathcal{N}_i^e} \textbf{Attention}(i, j) \cdot \textbf{Messaging}(i, j) \right), \tag{3.8}$$

where σ is implemented as $ReLU$.

Attention It augments the common self-attention mechanism to include relation features:

$$\textbf{Attention}(i, j) = softmax$$

$$\times \left(LeakyReLU \left(v^T \left[h_{e_i}^{in} || h_{e_j}^{in} || \frac{1}{|\mathcal{M}_{i,j}|} \sum_{r_k \in \mathcal{M}_{i,j}} h_{r_k} \right] \right) \right), \tag{3.9}$$

where $\mathcal{M}_{i,j}$ represents the set of linked relations that connect e_i to e_j. Noteworthily, it also adopts the multi-head attention mechanism to obtain the representation.

Messaging The features of neighboring entities are the corresponding features from the pre-processing stage.

Post-processing Finally, the outputs from different layers are concatenated to produce the final entity representations:

$$\hat{h}_{e_i}^{out} = \left[h_{e_i}^{out(0)} || \ldots || h_{e_i}^{out(l)} \right]. \tag{3.10}$$

Loss Function The loss function is formulated as:

$$\mathcal{L} = \sum_{(e_i, e_j) \in \mathcal{P}} ReLU(dis(e_i, e_j) - dis(e_i', e_j) + \lambda) + ReLU(dis(e_i, e_j)$$

$$- dis(e_i, e_j') + \lambda), \tag{3.11}$$

where $dis(\cdot, \cdot)$ is the Manhattan distance between two entity representations. e'_i and e'_j represent the negative instances.

3.2.3 RREA

It proposes to use relational reflection transformation to aggregate features for learning entity representations [9].

Aggregation The entity representations are denoted as:

$$h_{e_i}^{l+1} = ReLU \left(\sum_{e_j \in \mathcal{N}_{e_i}^e} \sum_{r_k \in \mathcal{R}_{ij}} \textbf{Attention}(i, j, k) \cdot \textbf{Messaging}(i, j, k) \right), \quad (3.12)$$

where $\mathcal{N}_{e_i}^e$ and $\mathcal{R}_{ij}$ represent the neighboring entity and relation sets, respectively.

Attention $\textbf{Attention}(i, j, k)$ denotes the weight coefficient computed by:

$$\textbf{Attention}(i, j, k) = \frac{\exp\left(\beta_{ijk}^l\right)}{\sum_{e_j \in \mathcal{N}_{e_i}^e} \sum_{r_k \in \mathcal{R}_{ij}} \exp\left(\beta_{ijk}^l\right)}, \quad (3.13)$$

where $\beta_{ijk}^l = \boldsymbol{v}^T[\boldsymbol{h}_{e_i}^l || \boldsymbol{M}_{r_k} \boldsymbol{h}_{e_j}^l || \boldsymbol{h}_{r_k}]$. $\boldsymbol{v}$ is a trainable vector. $\boldsymbol{M}_{r_k}$ is the relational reflection matrix of r_k. We leave out the details of relational reflection matrix in the interest of space, which can be found in the original paper.

Messaging The features of neighboring entities are the corresponding features from the pre-processing stage:

$$\textbf{Messaging}(i, j, k) = \boldsymbol{M}_{r_k} \boldsymbol{h}_{e_j}^l, \quad (3.14)$$

where $\boldsymbol{M}_{r_k}$ is the relational reflection matrix of r_k.

Post-processing Then, the outputs from different layers are concatenated to produce the output vector:

$$h_{e_i}^{out} = \left[\boldsymbol{h}_{e_i}^0 || \dots || \boldsymbol{h}_{e_i}^l \right]. \quad (3.15)$$

Finally, it concatenates the entity representation with its neighboring relation embeddings to obtain the final entity representation:

$$h_{e_i}^{Mul} = \left[h_{e_i}^{out} || \frac{1}{|\mathcal{N}_{e_i}^r|} \sum_{r_j \in \mathcal{N}_{e_i}^r} h_{r_j} \right]. \tag{3.16}$$

Loss Function The loss function is formulated as:

$$\mathcal{L} = \sum_{(e_i, e_j) \in \mathcal{P}} max\left(dis(e_i, e_j) - dis\left(e_i', e_j'\right) + \lambda, 0\right), \tag{3.17}$$

where $dis(\cdot, \cdot)$ is the Manhattan distance between two entity representations. e_i' and e_j' represent the negative instances generated by nearest neighbor sampling.

3.2.4 RPR-RHGT

This work introduces a meta path-based similarity framework for EA [2]. It considers the paths that frequently appear in the neighborhoods of pre-aligned entities to be reliable. We omit the generation of these reliable paths in the interest of space, which can be found in Sect. 3.3 of the original paper.

Pre-processing Specifically, it first generates relation embeddings by aggregating the representations of neighboring entities:

$$R^l(r) = \sigma \left[\frac{1}{|\mathcal{H}_r|} \sum_{e_i \in \mathcal{H}_r} b_h e_i^{l-1} || \frac{1}{|\mathcal{T}_r|} \sum_{e_j \in \mathcal{T}_r} b_t e_j^{l-1} \right], \tag{3.18}$$

where $\mathcal{H}_r$ and $\mathcal{T}_r$ denote the set of head entities and tail entities that are connected with relation r.

Aggregation The entity representation is obtained by averaging the messages from neighborhood entities with the attention weights:

$$\tilde{e}_h^l = \oplus_{\forall(r,t) \in RN(h)} H Attention(h, r, t) \cdot H Message(h, r, t), \tag{3.19}$$

where $\oplus$ denotes the overlay operation.

Attention The multi-head attention is computed as:

$$H Attention(h, r, t) = ||_{i \in [1, h_n]} softmax_{\forall(r,t) \in RN(h)}(H ATT_{head^i}(h, r, t)),$$

$$H ATT_{head^i}(h, r, t) = a^T ([K^i(h) || Q^i(t)] R^l(r)) / \sqrt{d/h_n}, \tag{3.20}$$

where $K^i(h) = K\_Linear^i(e_h^{l-1})$, $Q^i(t) = Q\_Linear^i(e_t^{l-1})$, $RN(h)$ represents the neighborhood entities of h, $\boldsymbol{a}$ denotes the learnable attention vector, h_n is the number of attention heads, and d/h_n is the dimension per head.

Messaging The multi-head message passing is computed as:

$$HMessage(h, r, t) = ||_{i \in [1,h_n]}(HMSG_{head^i}(h, r, t)),$$
$$HMSG_{head^i}(h, r, t) = [V\_Linear^i(e_t^{l-1})||R^l(r)], \tag{3.21}$$

where V_Linear^i is a linear projection of the tail entity, which is then concatenated with the relation representation.

Post-processing This work also combines the structural representations with name features using the residual connection:

$$e_h^l = \omega_\beta A\_Linear\left(\tilde{e}_h^l\right) + (1 - \omega_\beta)N\_Linear\left(e_h^{l-1}\right), \tag{3.22}$$

where A_Linear and N_Linear are linear projections. Correspondingly, based on the relation structure $\mathcal{T}_{rel}$ and path structure $\mathcal{T}_{path}$, it generates the relation-based embeddings $\boldsymbol{E}_{rel}$ and the path-based embeddings $\boldsymbol{E}_{path}$.

Loss Function Finally, the margin-based ranking loss function is used to formulate the overall loss function:

$$\mathcal{L} = \sum_{(p,q)\in\mathcal{L},(p',q')\in\mathcal{L}'_{rel}} [d_{rel}(p, q) - d_{rel}(p', q') + \lambda_1]_+$$
$$+\theta\left(\sum_{(p,q)\in\mathcal{L},(p',q')\in\mathcal{L}'_{path}} [d_{path}(p, q) - d_{path}(p', q') + \lambda_2]_+\right), \tag{3.23}$$

where the distance is measured by the Manhattan distance and θ is the hyper-parameter that controls the weights of relation loss and path loss.

3.2.5   RAGA

It proposes to adopt the self-attention mechanism to spread entity information to the relations and then aggregate relation information back to entities, which can further enhance the quality of entity representations [17].

Pre-processing In the pre-processing module, the pre-trained vectors are used as input and then forwarded to a two-layer GCN with highway network to encode structure information. We leave out the implementation details in the interest of space, which can be found in Sect. 4.2 in the original paper.

Aggregation In RAGA, there are three main GNN networks. Denote the initial representation of entity i as $\boldsymbol{h}_i$, which is generated in pre-processing module. The first GNN network obtains relation representation by aggregating all of its connected head entities and tail entities. For relation k, the aggregation of its connected head entities is computed as follows:

$$r_k^h = \sigma \left(\sum_{e_i \in \mathcal{H}_{r_k}} \sum_{e_j \in \mathcal{T}_{e_i r_k}} \textbf{Attention}_1(i, j, k) \cdot \textbf{Messaging}_1(i) \right), \qquad (3.24)$$

where σ is the ReLU activation function, $\mathcal{H}_{r_k}$ is the set of head entities for relation r_k, and $\mathcal{T}_{e_i r_k}$ is the set of tail entities for head entity e_i and relation r_k. The aggregation of all tail entities r_k^t can be computed through a similar process, and the relation representation is obtained as $r_k = r_k^h + r_k^t$.

Then, the second GNN network generates relation-aware entity representation through aggregating relation information back to entities. For entity i, the aggregation of all its outward relation embeddings is computed as follows:

$$h_i^h = \sigma \left(\sum_{e_j \in \mathcal{T}_{e_i}} \sum_{r_k \in \mathcal{R}_{e_i e_j}} \textbf{Attention}_2(i, k) \cdot r_k \right), \qquad (3.25)$$

where $\mathcal{T}_{e_i}$ is the set of *tail* entities for *head* entity e_i and $\mathcal{R}_{e_i e_j}$ is the set of relations between head entity e_i and tail entity e_j. The aggregation of inward relation embeddings h_i^t is computed through a similar process. Then the relation-aware entity representations h_i^{rel} can be obtained by concatenation: $h_i^{rel} = \left[h_i \| h_i^h \| h_i^t \right]$.

Finally, the third GNN takes as input the relation-aware entity representations and makes aggregation to produce the final entity representations:

$$h_i^{out} = \sigma \left(\sum_{j \in \mathcal{N}_i} \textbf{Attention}_3(i, j) \cdot h_k^{rel} \right), \qquad (3.26)$$

Attention Corresponding to three GNN networks, there are three attention computations in RAGA. In the first GNN, to compute the attention weights, representations of head entity and tail entity are linearly transformed, respectively, and then concatenated:

$$\textbf{Attention}_1(i, j, k) = \frac{\exp \left(\text{LeakReLU} \left(a_1^T \left[W^h h_i \| W^t h_j \right] \right) \right)}{\sum_{e_{i'} \in \mathcal{H}_{r_k}} \sum_{e_{j'} \in \mathcal{T}_{e_i r_k}} \exp \left(\text{LeakReLU} \left(a_1^T \left[W^h h_{i'} \| W^t h_{j'} \right] \right) \right)}, \qquad (3.27)$$

where a_1 is the learnable attention vector.

In the second GNN, representations of entity and its neighboring relations are directly concatenated:

$$\textbf{Attention}_2(i, k) = \frac{\exp\left(\text{LeakReLU}(a_2^T\,[h_i\,\|\,r_k])\right)}{\sum_{e_j \in T_{e_i}} \sum_{r_{k'} \in R_{e_i e_j}} \exp\left(\text{LeakReLU}(a_2^T\,[h_i\,\|\,r_{k'}])\right)},\qquad(3.28)$$

where a_2 is the learnable attention vector.

The computation of attention in the third GNN, i.e., **Attention**$_3$, is similar to Eq. (3.28), which concatenates entity and its neighboring entity instead of relation.

Messaging Only the first GNN utilizes linear transformation as the messaging approach:

$$\textbf{Messaging}_1(i) = W h_i\,,\qquad(3.29)$$

where W can refer to W^h or W^t depending on the aggregation of head or tail entities.

Post-processing The final enhanced entity representation is the concatenation of outputs of the second and the third GNNs:

$$h_i^{final} = \left[h_i^{rel}\,\|\,h_i^{out}\right].\qquad(3.30)$$

Loss Function The loss function is formulated as:

$$\mathcal{L} = \sum_{(e_i,e_j)\in T}\;\sum_{(e_i',e_j')\in T_{e_i,e_j}'} \max(dis(e_i, e_j) - dis(e_i', e_j') + \lambda, 0)\,,\qquad(3.31)$$

where T_{e_i,e_j}' is the set of negative sample for e_i and e_j, λ is the margin, and $dis()$ is defined as the Manhattan distance.

3.2.6 Dual-AMN

Dual-AMN proposes to utilize both intra-graph and cross-graph information for learning entity representations [7]. It constructs a set of virtual nodes, i.e., proxy vectors, through which the messaging and aggregation between graphs are conducted.

Aggregation Dual-AMN uses two GNN networks to learn intra-graph and cross-graph information, respectively. Firstly, it utilizes relation projection operation in RREA to obtain intra-graph embeddings:

$$h_{e_i}^l = \sigma\left(\sum_{e_j \in N_{e_i}}\;\sum_{r_k \in R_{ij}} \textbf{Attention}_1(i, j, k)\cdot\textbf{Messaging}_1(j, k)\right),\qquad(3.32)$$

where σ is the tanh activation function and $\boldsymbol{h}^l_{e_i}$ represents the output of l-th layer. Then the multi-hop embeddings are obtained by concatenation:

$$\boldsymbol{h}^{multi}_{e_i} = \left[\boldsymbol{h}^0_{e_i} \| \boldsymbol{h}^1_{e_i} \| \dots \| \boldsymbol{h}^l_{e_i} \right] . \tag{3.33}$$

Secondly, it constructs a set of virtual nodes $\mathcal{S}_p = \{\boldsymbol{q}_1, \boldsymbol{q}_2, \dots, \boldsymbol{q}_n\}$, namely, the proxy vectors, which are randomly initialized. The cross-graph aggregation is computed as:

$$\boldsymbol{h}^p_{e_i} = \sum_{j \in \mathcal{S}_p} \mathbf{Attention}_2(i, j) \cdot \mathbf{Messaging}_2(i, j) . \tag{3.34}$$

Attention For intra-graph information learning, the attention weights are calculated as:

$$\mathbf{Attention}_1(i, j, k) = \frac{\exp(\boldsymbol{v}^T \boldsymbol{h}_{r_k})}{\sum_{e_{j'} \in \mathcal{N}_{e_i}} \sum_{r_{k'} \in \mathcal{R}_{ij'}} \exp(\boldsymbol{v}^T \boldsymbol{h}_{r_{k'}})} , \tag{3.35}$$

where $\boldsymbol{v}^T$ is a learnable attention vector and $\boldsymbol{h}_{r_k}$ is the representation of relation r_k, which is randomly initialized by He_initializer [4].

For cross-graph information learning, the attention weights are computed by the similarity between entity and proxy vectors:

$$\mathbf{Attention}_2(i, j) = \frac{\exp(\cos(\boldsymbol{h}^{multi}_{e_i}, \boldsymbol{q}_j))}{\sum_{k \in \mathcal{S}_p} \exp(\cos(\boldsymbol{h}_{e_i}, \boldsymbol{q}_k))} . \tag{3.36}$$

Messaging For the first GNN, the messaging is the same as RREA, which utilizes a relational reflection matrix to transform neighbor embeddings.

For the second GNN, the features of neighboring entities are represented as the difference between entity and proxy vectors:

$$\mathbf{Messaging}_2(i, j) = \boldsymbol{h}^{multi}_{e_i} - \boldsymbol{q}_j . \tag{3.37}$$

Post-processing For the final entity embeddings, the gate mechanism is used to combine intra-graph and cross-graph representations:

$$\begin{aligned} \boldsymbol{\eta}_{e_i} &= \sigma(\boldsymbol{M}\boldsymbol{h}^p_{e_i} + \boldsymbol{b}), \\ \boldsymbol{h}^{final}_{e_i} &= \boldsymbol{\eta}_{e_i} \cdot \boldsymbol{h}^p_{e_i} + (1 - \boldsymbol{\eta}_{e_i}) \cdot \boldsymbol{h}^{multi}_{e_i} , \end{aligned} \tag{3.38}$$

where $\boldsymbol{M}$ and $\boldsymbol{b}$ are the gate weight matrix and gate bias vector.

Loss Function Firstly, it calculates the original margin loss as follows:

$$l_o(e_i, e_j, e'_j) = \gamma + \|\boldsymbol{h}_{e_i}^{final} - \boldsymbol{h}_{e_j}^{final}\|_2^2 - \|\boldsymbol{h}_{e_i}^{final} - \boldsymbol{h}_{e'_j}^{final}\|_2^2. \tag{3.39}$$

Inspired by batch normalization [5] which reduces the internal covariate shift, it proposes to use a normalization step that fixes the mean and variance of sample losses from $l_o(e_i, e_j, e'_j)$ to $l_n(e_i, e_j, e'_j)$ and reduces the dependence on the scale of the hyper-parameter. Finally, the overall loss function is defined as follows:

$$\mathcal{L} = \sum_{(e_i, e_j) \in P} \log \left[1 + \sum_{e'_j \in E_2} \exp(l_n(e_i, e_j, e'_j)) \right] \\ + \sum_{(e_i, e_j) \in P} \log \left[1 + \sum_{e'_i \in E_1} \exp(l_n(e_j, e_i, e'_i)) \right], \tag{3.40}$$

where P is the set of positive samples and E_1 and E_2 are the sets of entities in two knowledge graphs, respectively.

3.2.7 ERMC

This work proposes to jointly model and align entities and relations and meanwhile retain their semantic independence [14].

Pre-processing For pre-processing, it obtains names or descriptions of entities and relations as the inputs for BERT [6] and adds an MLP layer to construct initial representations, which are denoted as $\boldsymbol{x}^{e(0)}$ and $\boldsymbol{x}^{r(0)}$ for each entity and relation, respectively.

Aggregation Given an entity e, the model first aggregates the embeddings of entities that point to e:

$$\boldsymbol{h}_{\mathcal{N}_i^e}^{e(l+1)} = \sigma \left(\frac{1}{|\mathcal{N}_i^{e(e)}|} \sum_{e_i \in \mathcal{N}_i^{e(e)}} \textbf{Messaging}(i) \right), \tag{3.41}$$

where $\sigma(\cdot)$ contains normalization, dropout, and activation operations. Similarly, the model aggregates the embeddings of entities that e points to, the embeddings of relations that point to e, and the embeddings of relations that e points to, producing $\boldsymbol{h}_{\mathcal{N}_i^r}^{e(l+1)}$, $\boldsymbol{h}_{\mathcal{N}_o^e}^{e(l+1)}$, and $\boldsymbol{h}_{\mathcal{N}_o^r}^{e(l+1)}$, respectively. The model also aggregates the embeddings of entities that point to a relation r or r points to, so as to produce the relation embeddings $h_{\mathcal{N}_i^e}^{r(l+1)}$ and $h_{\mathcal{N}_o^e}^{r(l+1)}$, respectively.

Messaging Given an entity e, the messaging process of the entities that point to e is implemented as a simple linear transformation: $\textbf{Messaging}(i) = W_{e_i}^{e(l)} x^{e_i(l)}$, where $x^{e_i(l)}$ is the node representation in the last layer and $W_{e_i}^{e(l)}$ is a learnable weight matrix that aggregates the inward entity features. The messaging process of other operations is implemented similarly.

Post-processing The final representation of entity e is formulated as follows:

$$h^{e(l+1)} = \left[h_{\mathcal{N}_i^e}^{e(l+1)} \| h_{\mathcal{N}_i^r}^{e(l+1)} \| h_{\mathcal{N}_o^e}^{e(l+1)} \| h_{\mathcal{N}_o^r}^{e(l+1)} \right],$$

$$x^{e(l+1)} = MLP \left(\left[h^{e(l+1)} \| x^{e(l)} \right] \right) . \tag{3.42}$$

And the final representation of relation r is formulated similarly:

$$h^{r(l+1)} = \left[h_{\mathcal{N}_i^e}^{r(l+1)} \| h_{\mathcal{N}_o^e}^{r(l+1)} \right],$$

$$x^{r(l+1)} = MLP \left(\left[h^{r(l+1)} \| x^{r(l)} \right] \right) . \tag{3.43}$$

The graph embedding $H \in \mathbb{R}^{(|E|+|R|) \times d}$ is the concatenation of all entities and relations' representations.

Loss Function Denote H_s and H_t as the representations of two graphs, respectively. The similarity matrix is computed as:

$$S = sinkhorn(H_s, H_t^T), \tag{3.44}$$

where $s_{i,j} \in S$ is a real number that denotes the correlation between entity e_s^i (from source graph) and e_t^j (from target graph), or the correlation between relation r_s^i (from source graph) and r_t^j (from target graph). The other elements are set to $-\infty$ to mask the correlation between entity and relation across different graphs. The final loss function is formulated as follows:

$$\mathcal{L} = - \sum_{\left(e_s^i, e_t^j \right) \in Q^e} \log \left(s_{i,j} \right) - \lambda \sum_{\left(r_s^i, r_t^j \right) \in Q^r} \log \left(s_{i,j} \right) , \tag{3.45}$$

where (e_s^i, e_t^j) and (r_s^i, r_t^j) are pre-aligned entity and relation pairs and $\lambda \in [0, 1]$ is a hyper-parameter.

3.2.8 KE-GCN

It combines GCNs and advanced KGE methods to learn the representations, where a novel framework is put forward to realize the messaging and aggregation modules in representation learning [15].

Aggregation Denoting h_v^l as the embedding of entity v at layer l, the entity updating rules are:

$$m_v^{l+1} = \sum_{(u,r)\in\mathcal{N}_{\text{in}}(v)} \textbf{Messaging}(u, r, v) + \sum_{(u,r)\in\mathcal{N}_{\text{out}}(v)} \textbf{Messaging}(u, r, v),$$

$$h_v^{l+1} = \sigma(m_v^{l+1} + W_0^l h_v^l),$$

$$(3.46)$$

where $\mathcal{N}_{\text{in}}(v) = \{(u, r)|u \xrightarrow{r} v\}$ is the set of inward entity-relation neighbors of entity v, while $\mathcal{N}_{\text{out}}(v) = \{(u, r)|u \xleftarrow{r} v\}$ is the set of outward neighbors of v. W_0^l is a linear transformation matrix. $\sigma(\cdot)$ denotes the activation function for the update. The embedding of relation is updated through a similar process.

Messaging It considers GCN as an optimization process, where the messaging process is implemented as a partial derivative:

$$\textbf{Messaging}(u, r, v) = W_r^l \frac{\partial f(h_u^l, h_r^l, h_v^l)}{\partial h_v^l}, \qquad (3.47)$$

where h_r^l represents the embedding of relation r at layer l and W_r^l is a relation-specific linear transformation matrix. $f(h_u^l, h_r^l, h_v^l)$ is the scoring function that measures the plausibility of triple (u, r, v). Thus, $m_v^{l+1} + W_0^l h_v^l$ in Eq. (3.46) can be regarded as the gradient ascent to maximize the sum of scoring function. For example, if $f(h_u^l, h_r^l, h_v^l) = (h_u^l)^T h_v^l$, Eq. (3.47) becomes equivalent to the common linear transformation $W_r^l h_u^l$.

Loss Function Denote the training set as $S = \{(u, v)\}$; this model utilizes margin-based ranking loss for optimization:

$$\mathcal{L} = \sum_{(u,v)\in S} \sum_{(u',v')\in S'_{(u,v)}} \max(\|h_u - h_v\|_1 + \gamma - \|h_{u'} - h_{v'}\|_1, 0), \qquad (3.48)$$

where $S'_{(u,v)}$ denotes the set of negative entity alignments constructed by corrupting (u, v), i.e., replacing u or v with a randomly chosen entity in graph. γ represents the margin hyper-parameter separating positive and negative entity alignments.

3.2.9 RePS

It encodes position and relation information for aligning entities [13].

Aggregation Firstly, to encode position information, k subsets of nodes (referred to as anchor sets) are randomly sampled. An ith anchor set is a collection of l_i number of nodes (anchors). Then for entity v, the aggregation process is formulated as:

$$h_{v_p}^l = g \left(\frac{1}{k+1} \left(\sum_{i=1}^{k} \text{Messaging}_1(v, \psi_i) + h_v^{l-1} \right) \right), \tag{3.49}$$

where h_v^l represents the embedding of entity v from layer l, ψ_i is the ith anchor set, and $g(X) = \sigma(W_1 X + b_1)$, where W_1 and b_1 are trainable parameters and σ is the activation function.

To encode relation information, a simple relation-specific GNN is used:

$$h_{v_r}^l = f \left((1 + c_v) \cdot h_v^{l-1} + \sum_{i \in \mathcal{N}_v} \text{Messaging}_2(i) \right), \tag{3.50}$$

where c_v is the learnable coefficient for entity v and $\mathcal{N}_v$ is the set of neighboring entities of v. $f(X) = W_2 X + b_2$, where W_2 and b_2 are learnable parameters.

Messaging To ensure similar entities in two graphs have similar representations, the relation-enriched distance function is defined as follows:

$$pd(u, v) = \min_q \left(\sum_{r \in P_q(u,v)} f(r, \mathcal{KG}_i) \right), \tag{3.51}$$

where $f(r, \mathcal{KG}_i)$ is the frequency of relation r in $\mathcal{KG}_i$ and $P_q(u, v)$ is the list of relations in the qth path between u and v. Thus, $pd(u, v)$ aims to find the shortest path between u and v, where the relations appear less frequently. Then the messaging function is formulated as follows:

$$\text{Messaging}_1(v, \psi_i) = \min \left(\left\{ pd(v, \phi_{i,j}) \cdot h_{\psi_{i,j}}^{l-1} \right\}_{j=1}^{l_i} \right), \tag{3.52}$$

where $\psi_{i,j}$ is the jth entity in ith anchor set.

For relation-aware embedding, it sums up the neighboring representations with relation-specific weights:

$$\text{Messaging}_2(i) = \frac{h_i^{l-1}}{1 + c_{r_{v,i}}}, \tag{3.53}$$

where $c_{r_{v,i}}$ is the learnable coefficient for relation r connecting v and i.

Post-processing The final representation of v is computed as:

$$h_v^l = g\left(h_{v_p}^l\right) \cdot h_{v_p}^l + \left(1 - g\left(h_{v_p}^l\right)\right) \cdot h_{v_r}^l, \tag{3.54}$$

where $g(h_{v_p}^l) = \sigma(W_3 h_{v_p}^l + b_3)$ learns the relative importance. W_3 and b_3 are trainable parameters and σ is the activation function.

Loss Function It introduces a novel knowledge-aware negative sampling (KANS) technique to generate hard negative samples. For each tuple (v, v') in S, the negative instances for v are sampled from set Φ_v, where Φ_v is the set of entities which share at least one (relation, tail) pair or (relation, head) pair with v'. The model is trained by minimizing the following loss:

$$\mathcal{L} = \sum_{(p,p')\in S} \|p - p'\| + \beta \sum_{(p,q)\in S'} [\gamma - \|p - q\|]_+, \tag{3.55}$$

where β is a weighing parameter and γ is the margin.

3.2.10 SDEA

SDEA utilizes BiGRU to capture correlations among neighbors and generate entity representations [16].

Pre-processing It devises an attribute embedding module to capture entity associations via entity attributes. Specifically, given an entity e_i, it concatenates the names and descriptions of its attributes, denoted as $S(e_i)$. Then $S(e_i)$ is fed into BERT model to generate attribute embedding $H_a(e_i)$. The details of implementation can be found in Section III of the original paper, which is omitted in the interest of space.

Aggregation It aggregates the neighboring information utilizing attention mechanism:

$$H_r(e_i) = \sum_{t=1}^{n} \text{Attention}(t) \cdot \text{Messaging}(t). \tag{3.56}$$

Since SDEA treats neighborhood as a sequence, t actually represents t-th neighboring entity of e_i, and **Messaging**() is computed through a BiGRU.

Attention SDEA computes attention via simple inner product:

$$\textbf{Attention}(t) = \frac{\exp\left(h_t^T \cdot \hat{h}\right)}{\sum_{i=1}^{n} \exp\left(h_i^T \cdot \hat{h}\right)}, \tag{3.57}$$

where $\hat{h}$ is the global attention representation, which is obtained after feeding the output of the last unit of the BiGRU, denoted as h_n, into an MLP layer.

Messaging Different from other models, SDEA captures correlation between neighbors in messaging module, and all neighbors of entity e_i are regarded as an input sequence of the BiGRU model. Given entity e_i, let x_t denote the t-th input embedding (i.e., the attribute embedding of e_i's t-th neighbor, as described in pre-processing module) and h_t denote the output t-th hidden unit. The process of BiGRU is formulated as follows:

$$
\begin{aligned}
r_t &= \sigma(W_r x_t + U_r h_{t-1} + b_r) \\
\tilde{h}_t &= \phi(W x_t) + U(r_t \odot h_{t-1} + b_h) \\
z_t &= \sigma(W_z x_t + U_z h_{t-1} + b_z) \\
h_t &= (1 - z_t) \odot h_{t-1} + z_t \odot \tilde{h}_t,
\end{aligned}
\tag{3.58}
$$

where r_t is the reset gate that drops the unimportant information and z_t is the update gate that combines the important information. W, U, b are learnable parameters. $\tilde{h}_t$ is the hidden state. σ is the sigmoid function and ϕ is the hyperbolic tangent. $\odot$ is the Hadamard product.

For BiGRU, there are outputs of two directions $\overleftarrow{h}_t$ and $\overrightarrow{h}_t$, and the final output of BiGRU, namely, the output of messaging module, is the sum of two directions: **Messaging**$(i) = \overleftarrow{h}_t + \overrightarrow{h}_t$.

Post-processing After obtaining the attribute embedding $H_a(e_i)$ and the relational embedding $H_r(e_i)$, they are concatenated and forwarded to another MLP layer, resulting in $H_m(e_i) = MLP([H_a(e_i) \| H_r(e_i)])$. Finally, $H_a(e_i)$, $H_r(e_i)$, and $H_m(e_i)$ are concatenated to produce $H_{ent}(e_i) = [H_r(e_i) \| H_a(e_i) \| H_m(e_i)]$, which is used in alignment stage.

Loss Function The model uses the following margin-based ranking loss as the loss function to train attribute embedding module:

$$\mathcal{L} = \sum_{e_i, e_i', e_i'' \in D} \max\left\{0, \|H_a(e_i) - H_a'(e_i')\|_2 - \|H_a(e_i) - H_a'(e_i'')\|_2 + \beta\right\},$$

$$\tag{3.59}$$

where D is the training set; $\boldsymbol{H}_a$ and $\boldsymbol{H}'_a$ are attribute embeddings of source graph and target graph, respectively; and $\beta > 0$ is the margin hyper-parameter used for separating positive and negative pairs.

The training of relation embedding module uses a margin-based ranking loss similar to Eq. (3.59), where the embedding $\boldsymbol{H}_a(e_i)$ is replaced by $[\boldsymbol{H}_r(e_i) \| \boldsymbol{H}_m(e_i)]$.

3.3 Experiments

In this section, we first conduct overall comparison experiment to reveal the effectiveness of state-of-the-art representation learning methods. Then we conduct further experiments in terms of the six modules of representation learning, so as to examine the effectiveness of various strategies.

3.3.1 Experimental Setting

Dataset We use the most frequently used DBP15K dataset [11] for evaluation.

Baselines For overall comparison, we select seven models, including AliNet [12], MRAEA [8], RREA [9], RAGA [17], SDEA [16], Dual-AMN [7], and RPR-RHGT [2]. We collect their source codes and reproduce the results in the same setting. Specifically, to make a fair comparison, we modify and unify the alignment part of these models, forcing them to utilize L1 distance and greedy algorithm for alignment inference. We omit the comparison with the remaining models, as they do not provide the source codes and our implementations cannot reproduce the results. For ablation and further experiments, we choose RAGA as the base model.

Parameters and Metrics Since there are various kinds of hyper-parameters for different models, we just unify the common parameters, such as the margin $\lambda = 3$ in margin loss function, and number of negative samples $k = 5$. For other parameters, we keep the default settings in the original papers.

Following existing studies, we use Hits@k ($k = 1, 10$) and mean reciprocal rank (MRR) as the evaluation metrics. The higher the Hits@k and MRR, the better the performance. In experiments, we report the average performance of three independent runs as the final result.

3.3.2 Overall Results and Analysis

Firstly, we compare the overall performance of seven advanced models in Table 3.2, where the best results are highlighted in bold, and the second best results are underlined.

Table 3.2 Comparison of representation learning models on DBP15K

Model	ZH-EN			JA-EN			FR-EN		
	Hits@1	Hits@10	MRR	Hits@1	Hits@10	MRR	Hits@1	Hits@10	MRR
AliNet	51.7	80.9	0.616	52.1	80.3	0.619	52.1	83.5	0.628
MRAEA	64.2	89.8	0.733	65.4	89.7	0.743	67.6	92.3	0.767
RREA	56.8	87.3	0.672	57.0	87.6	0.675	59.3	90.0	0.700
RAGA	77.8	92.3	0.831	81.8	94.6	0.866	91.0	98.1	0.937
SDEA	**84.3**	**95.0**	**0.883**	79.2	90.7	0.833	94.8	98.7	0.964
Dual-AMN	79.6	92.7	0.843	79.0	93.1	0.842	82.1	95.0	0.870
RPR-RHGT	65.0	84.9	0.720	**87.3**	**95.1**	**0.902**	86.1	95.5	0.895

From the results, it can be observed that:

- No model achieves state-of-the-art performance over all three KG pairs. This indicates that current advanced models have advantages and disadvantages in different situations.
- SDEA achieves the best performance on ZH-EN and FR-EN, and RPR-RHGT leads on JA-EN. Considering that both of the two models leverage pre-trained model to obtain initial embeddings and devise novel approaches to extract neighboring features, we may draw primary conclusion that utilizing pre-trained model benefits the representation learning process, and effective messaging approach is important to the overall results.
- RAGA achieves the second best performance on JA-EN and FR-EN, and Dual-AMN attains the second best result on ZH-EN. Notably, RAGA also leverages the pre-trained model, which further validates the effectiveness of using pre-trained model for initialization. Dual-AMN uses proxy vectors that can help capture cross-graph information and hence improve representation learning.
- AliNet performs the worst over three datasets. As AliNet is the only model that aggregates 2-hop neighboring entities, it may indicate directly incorporating 2-hop neighboring information benefits little, which can also be observed in the further experiments on aggregation module.

3.3.3 Further Experiments

To compare various strategies in each module of representation learning, we conduct further experiments using the RAGA model.

3.3.3.1 Pre-processing Module

RAGA takes pre-trained embeddings as input, which are forwarded to a two-layer GCN with highway network to generate initial representations. To examine the

Table 3.3 Analysis of the pre-processing module using RAGA

Model	ZH-EN			JA-EN			FR-EN		
	Hits@1	Hits@10	MRR	Hits@1	Hits@10	MRR	Hits@1	Hits@10	MRR
RAGA	**77.7**	**92.3**	**0.831**	**81.8**	**94.6**	**0.866**	**91.0**	**98.1**	**0.937**
RAGA w/o Pre-trained	34.6	58.4	0.430	34.9	58.4	0.433	33.9	59.0	0.427
RAGA w/o GNN	<u>72.4</u>	<u>85.5</u>	<u>0.772</u>	<u>77.8</u>	<u>88.8</u>	<u>0.818</u>	<u>89.9</u>	<u>96.1</u>	<u>0.923</u>
RAGA w/o Both	19.8	34.5	0.250	20.3	36.3	0.260	20.1	36.7	0.259

effectiveness of pre-trained embeddings and structural embeddings, we remove them, respectively, and then make comparison. Table 3.3 shows the results, where "w/o Pre-trained" represents removing pre-trained embeddings, "w/o GNN" represents removing GCN, and "w/o Both" represents removing the whole pre-processing module.

The results show that removing the structural features and the pre-trained embeddings significantly degrades the performance, and the model that completely removes the pre-processing module achieves the worst result. Hence, it is important to extract useful features to initialize the embeddings. Additionally, we can also observe that the semantic features in the pre-trained model are more useful than the structural vectors, which verifies the effectiveness of the prior knowledge contained in the pre-trained embeddings. Using structural embeddings for initialization is less effective, as the subsequent steps in representation learning also aim to extract the structural features to produce meaningful representations.

3.3.3.2   Messaging Module

For the messaging module, linear transformation is the most widely used approach. RAGA only utilizes linear transformation in its first GNN and does not use transformation in the other two GNNs. Thus, we design two variants: one that eliminates the linear transformation in the first GNN ("-Linear Transform"), resulting in a model without linear transformation at all, and the other one that adds linear transformation in the other two GNNs ("+Linear Transform"), resulting in a model that is fully equipped with linear transformation.

The results are presented in Table 3.4. Besides, we also report their convergence rates in Fig. 3.1.

It is evident that adding linear transformation in the rest of the GNNs improves the performance of RAGA, especially on JA-EN and FR-EN datasets, where Hits@1 improves by 1.1% and 1.2%, respectively. Additionally, when removing linear transformation, the performance drops significantly. Furthermore, Fig. 3.1 shows that linear transformation can also boost the convergence of model, possibly due to the introduction of extra parameters.

Table 3.4 Analysis of the messaging module using RAGA

Model	ZH-EN			JA-EN			FR-EN		
	Hits@1	Hits@10	MRR	Hits@1	Hits@10	MRR	Hits@1	Hits@10	MRR
RAGA	77.7	**92.3**	0.831	81.8	94.6	0.866	91.0	98.1	0.937
RAGA +Linear Transform	**78.3**	**92.3**	**0.834**	**82.9**	**95.0**	**0.873**	**92.2**	**98.5**	**0.946**
RAGA −Linear Transform	68.3	85.1	0.744	70.8	86.2	0.763	81.6	92.3	0.855

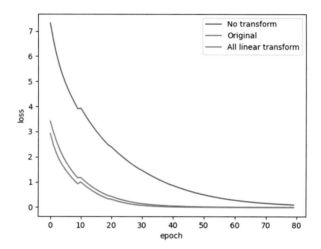

Fig. 3.1 Comparison of convergences

3.3.3.3 Attention Module

For attention module, there are two popular implementations, i.e., inner product and concatenation. To compare the two approaches, we replace the concatenation computation of RAGA with inner product computation (i.e., "-Inner product," by changing $v^T[e_i \| e_j]$ to $(M_1 e_i)^T(M_2 e_j)$, where M_1, M_2 are learnable transformation matrices), and remove the attention mechanism, (i.e., "w/o Attention," where we do not compute attention coefficient and just take average operation), respectively, and then report the results.

As it is shown in Table 3.5, the two variant models perform almost the same as the original model. Considering the influence of the initial representation generated in the pre-processing module, we remove the pre-trained vectors of the pre-processing module and then conduct the same comparison. As shown in Table 3.6, removing the attention mechanism drops the performance, so we may draw a preliminary conclusion that the attention mechanism can play a better role in the absence of prior knowledge. As for the two strategies of attention computation, inner product

Table 3.5 Analysis of the attention module using RAGA

Model	ZH-EN			JA-EN			FR-EN		
	Hits@1	Hits@10	MRR	Hits@1	Hits@10	MRR	Hits@1	Hits@10	MRR
RAGA	77.7	92.3	0.831	81.8	94.6	0.866	91.0	98.1	0.937
RAGA-Inner product	77.8	92.2	0.831	81.8	94.6	0.866	91.0	98.1	0.937
RAGA w/o Attention	77.7	92.3	0.831	81.8	94.6	0.865	91.0	98.1	0.937

Table 3.6 Analysis of the attention module using RAGA after removing pre-trained embeddings

Model	ZH-EN			JA-EN			FR-EN		
	Hits@1	Hits@10	MRR	Hits@1	Hits@10	MRR	Hits@1	Hits@10	MRR
RAGA	34.6	58.4	0.430	**34.9**	**58.4**	**0.433**	33.9	**59.0**	**0.427**
RAGA-Inner product	**35.1**	**59.0**	**0.435**	34.6	58.2	0.430	33.4	58.5	0.422
RAGA w/o Attention	34.5	58.3	0.430	33.9	57.4	0.422	33.0	58.1	0.418

Table 3.7 Analysis of the aggregation module using RAGA

Model	ZH-EN			JA-EN			FR-EN		
	Hits@1	Hits@10	MRR	Hits@1	Hits@10	MRR	Hits@1	Hits@10	MRR
RAGA	**77.7**	**92.3**	**0.831**	**81.8**	**94.6**	**0.866**	**91.0**	**98.1**	**0.937**
RAGA-2hop	77.3	92.2	0.827	81.5	94.5	0.862	91.0	98.1	0.937
RAGA w/o rel.	50.0	77.0	0.597	52.5	78.6	0.619	54.2	79.4	0.631

performs better than concatenation on ZH-EN dataset but worse on JA-EN and FR-EN datasets, which indicates these two approaches make different contributions on different datasets.

3.3.3.4 Aggregation Module

For the aggregation module, as RAGA incorporates both 1-hop neighbors and relation information to update entity representations, we examine two variants, i.e., adding two hop neighboring information ("-2hop") and removing relation representation ("w/o rel."). The results are shown in Table 3.7.

We can observe that the performance of the model decreases significantly after removing the relation representation learning. This shows that the integration of relation representations can indeed enhance the learning ability of the model. Besides, the performance of the model decreases slightly after adding the information of 2-hop neighboring entities, which might indicate that the 2-hop neighboring information can bring some noises, as not all entities are useful for aligning the target entity.

Table 3.8 Analysis of the post-processing module using RAGA

Model	ZH-EN			JA-EN			FR-EN		
	Hits@1	Hits@10	MRR	Hits@1	Hits@10	MRR	Hits@1	Hits@10	MRR
RAGA	**77.7**	92.3	0.831	**81.8**	**94.6**	**0.866**	91.0	98.1	0.937
RAGA-highway	77.6	**92.7**	**0.832**	81.2	94.2	0.861	**91.1**	**98.3**	**0.939**
RAGA w/o post-processing	52.7	81.7	0.629	52.4	81.8	0.628	55.9	85.0	0.661

Table 3.9 Analysis of the loss function module using RAGA

Model	ZH-EN			JA-EN			FR-EN		
	Hits@1	Hits@10	MRR	Hits@1	Hits@10	MRR	Hits@1	Hits@10	MRR
RAGA	**77.7**	**92.3**	**0.831**	**81.8**	**94.6**	**0.866**	**91.0**	**98.1**	**0.937**
RAGA-TransE	50.5	68.5	0.569	53.1	74.8	0.605	71.9	85.1	0.766
RAGA w/ TransE	73.1	89.0	0.789	77.8	92.1	0.831	88.3	96.5	0.915

3.3.3.5 Post-processing Module

RAGA concatenates the relation-aware entity representation and the 1-hop aggregation results to produce the final representation. We examine two variants, i.e., "-highway' 'that replaces concatenation with highway network [10], and "w/o post-processing" that removes the relation-aware entity representation (Table 3.8).

From the experimental results, it can be seen that removing post-processing module decreases the performance, which indicates that the relation-aware representations can indeed enhance the final representations and improve the alignment performance. After replacing the concatenation operation with highway network, the performance decreases on JA-EN dataset and increases on FR-EN dataset, which indicates that the two post-processing strategies do not have absolute advantages and disadvantages.

3.3.3.6 Loss Function Module

For the loss function, RAGA employs margin-based loss in training. We consider two other popular choices, i.e., TransE-based loss and margin-based + TransE loss. Specifically, TransE-based loss is formulated as $l_E = \frac{1}{k}\sum_k \|h_k + r_k - t_k\|_1$, where (h_k, r_k, t_k) is randomly sampled.

From the results in Table 3.9, it can be seen that the model performance decreases after using or adding the TransE loss. This is mainly because the TransE assumption is not universal. For example, in the RAGA model used in this experiment, the representation of the relation is actually obtained by adding the head entity and the tail entity, which is in conflict with the TransE assumption.

3.4   Conclusion

In this chapter, we survey recent advance in the representation learning stage of EA. We propose a general framework of GNN-based representation learning models, which consists of six modules, and summarize ten recent works in terms of these modules. Extensive experiments are conducted to show the overall performance of each method and also reveal the effectiveness of the strategies in each module.

References

1. Antoine Bordes, Nicolas Usunier, Alberto Garcia-Duran, Jason Weston, and Oksana Yakhnenko. Translating embeddings for modeling multi-relational data. *NIPS*, 26, 2013.
2. Weishan Cai, Wenjun Ma, Jieyu Zhan, and Yuncheng Jiang. Entity alignment with reliable path reasoning and relation-aware heterogeneous graph transformer. In Lud De Raedt, editor, *Proceedings of the Thirty-First International Joint Conference on Artificial Intelligence, IJCAI-22*, pages 1930–1937. International Joint Conferences on Artificial Intelligence Organization, 7 2022. Main Track.
3. Marco Cuturi. Sinkhorn distances: Lightspeed computation of optimal transport. *NIPS*, 26, 2013.
4. Kaiming He, Xiangyu Zhang, Shaoqing Ren, and Jian Sun. Delving deep into rectifiers: Surpassing human-level performance on imagenet classification. In *ICCV*, pages 1026–1034, 2015.
5. Sergey Ioffe and Christian Szegedy. Batch normalization: Accelerating deep network training by reducing internal covariate shift. In *ICML*, pages 448–456. PMLR, 2015.
6. Jacob Devlin Ming-Wei Chang Kenton and Lee Kristina Toutanova. Bert: Pre-training of deep bidirectional transformers for language understanding. In *Proceedings of NAACL-HLT*, pages 4171–4186, 2019.
7. Xin Mao, Wenting Wang, Yuanbin Wu, and Man Lan. Boosting the speed of entity alignment 10×: Dual attention matching network with normalized hard sample mining. In *WWW 2021*, pages 821–832, 2021.
8. Xin Mao, Wenting Wang, Huimin Xu, Man Lan, and Yuanbin Wu. Mraea: an efficient and robust entity alignment approach for cross-lingual knowledge graph. In *WSDM*, pages 420–428, 2020.
9. Xin Mao, Wenting Wang, Huimin Xu, Yuanbin Wu, and Man Lan. Relational reflection entity alignment. In *CIKM*, pages 1095–1104, 2020.
10. Rupesh Kumar Srivastava, Klaus Greff, and Jürgen Schmidhuber. Highway networks. *arXiv preprint arXiv:1505.00387*, 2015.
11. Zequn Sun, Wei Hu, and Chengkai Li. Cross-lingual entity alignment via joint attribute-preserving embedding. In *ISWC(1)*, pages 628–644. Springer, 2017.
12. Zequn Sun, Chengming Wang, Wei Hu, Muhao Chen, Jian Dai, Wei Zhang, and Yuzhong Qu. Knowledge graph alignment network with gated multi-hop neighborhood aggregation. In *AAAI*, volume 34, pages 222–229, 2020.
13. Anil Surisetty, Deepak Chaurasiya, Nitish Kumar, Alok Singh, Gaurav Dhama, Aakarsh Malhotra, Ankur Arora, and Vikrant Dey. Reps: Relation, position and structure aware entity alignment. In *WWW 2022*, pages 1083–1091, 2022.
14. Jinzhu Yang, Ding Wang, Wei Zhou, Wanhui Qian, Xin Wang, Jizhong Han, and Songlin Hu. Entity and relation matching consensus for entity alignment. In *CIKM*, pages 2331–2341, 2021.
15. Donghan Yu, Yiming Yang, Ruohong Zhang, and Yuexin Wu. Knowledge embedding based graph convolutional network. In *WWW*, pages 1619–1628, 2021.

16. Ziyue Zhong, Meihui Zhang, Ju Fan, and Chenxiao Dou. Semantics driven embedding learning for effective entity alignment. In *2022 IEEE 38th International Conference on Data Engineering (ICDE)*, pages 2127–2140. IEEE, 2022.
17. Renbo Zhu, Meng Ma, and Ping Wang. Raga: relation-aware graph attention networks for global entity alignment. In *Advances in Knowledge Discovery and Data Mining: 25th Pacific-Asia Conference, PAKDD 2021, Virtual Event, May 11–14, 2021, Proceedings, Part I*, pages 501–513. Springer, 2021.

Chapter 4
Recent Advance of Alignment Inference Stage

Abstract In this chapter, we introduce recent progress of the alignment inference stage.

4.1 Introduction

Matching data instances that refer to the same real-world entity is a long-standing problem. It establishes the connections among multiple data sources and is critical to data integration and cleaning [33]. Therefore, the task has been actively studied; for instance, in the database community, various entity matching (EM) (and entity resolution (ER)) strategies are proposed to train a (supervised) classifier to predict whether a pair of data records match [9, 33].

Recently, due to the emergence and proliferation of knowledge graphs (KGs), matching entities in KGs draws much attention from both academia and industries. Distinct from traditional data matching, it brings its own challenges. In particular, it underlines the use of KGs' structures for matching and manifests unique characteristics of data, e.g., imbalanced class distribution, and few attributive textual information, etc. As a consequence, although viable, following the traditional EM pipeline, it is difficult to train an effective classifier that can infer the equivalence between KGs' entities. Hence, much effort has been dedicated to specifically addressing the matching of entities in KGs, which is also referred to as *entity alignment* (EA).

Nevertheless, early solutions to EA are mainly unsupervised [22, 42], i.e., no labeled data is assumed. They utilize discriminative features of entities (e.g., entity descriptions and relational structures) to infer the equivalent entity pair, which are, however, embarrassed by the heterogeneity of independently constructed KGs [44].

To mitigate this issue, recent solutions to EA employ a few labeled pairs as seeds to guide the learning and prediction [8, 14, 27, 37, 47]. In short, they embed the symbolic representations of KGs as low-dimensional vectors in a way such that the semantic relatedness of entities is captured by the geometrical structures of embedding spaces [4], where the seed pairs are leveraged to produce unified entity representations. In the testing stage, they match entities based on the unified

X. Zhao et al., *Entity Alignment*, Big Data Management,
https://doi.org/10.1007/978-981-99-4250-3_4

entity embeddings. They are coined as *embedding-based* EA methods, which have exhibited state-of-the-art performance on existing benchmarks.

To be more specific, the embedding-based EA[1] pipeline can be roughly divided into two major stages, i.e., *representation learning* and *matching KGs in entity embedding spaces* (or *embedding matching* for short). While the former encodes the KG structures into low-dimensional vectors and establishes connections between independent KGs via the calibration or transformation of (seed) *entity embeddings* [44], the latter computes pairwise scores between source and target entities based on such embeddings and then makes alignment decisions according to the pairwise scores. Although this field has been actively explored, existing efforts are mainly devoted to the *representation learning* stage [24, 26, 61], while *embedding matching* has not raised many attentions until very recently [29, 53]. The majority of existing EA solutions adopt a simple algorithm to realize this stage, i.e., DInf, which first leverages common similarity metrics such as cosine similarity to calculate the pairwise similarity scores between entity embeddings and then matches a source entity to its most similar target entity according to the pairwise scores [47]. Nevertheless, it is evident that such an intuitive strategy can merely reach local optimums for individual entities and completely overlooks the (global) interdependence among the matching decisions for different entities [55].

To address the shortcomings of DInf, advanced strategies are devised [12, 44, 49, 53, 55, 56]. While some of them inject the modeling of global interdependence into the computation of pairwise scores [12, 44, 53], some directly improve the alignment decision-making process by imposing collective matching constraints [49, 55, 56]. These efforts demonstrate the significance of *matching KGs in entity embedding spaces* from at least three major aspects: (1) It is an indispensable step of EA, which takes as input the entity embeddings (generated by the representation learning stage), and outputs matched entity pairs. (2) Its performance is crucial to the overall EA results, e.g., an effective algorithm can improve the alignment results by up to 88% [53]. (3) It empowers EA with explainability, as it unveils the decision-making process of alignment. We use the following example to further illustrate the significance of the embedding matching process.

Example Figure 4.1 presents three representative cases of EA. The KG pairs to be aligned are first encoded into embeddings via the representation learning models. Next, the embedding matching algorithms produce the matched entity pairs based on the embeddings. In the most ideal case where two KGs are identical, e.g., case (a), with an ideal representation learning model, equivalent entities would be embedded into exactly the same place in the

(continued)

---

[1] In the rest of the paper, we use *EA* to refer to embedding-based EA solutions and *conventional EA* for the early solutions.

low-dimensional space, and using the simple DInf algorithm would attain perfect results. Nevertheless, in the majority of practical scenarios, e.g., case (b) and (c), the two KGs have high structure heterogeneity. As thus, even an ideal representation learning model might generate different embeddings for equivalent entities. In this case, adopting the simple DInf strategy is likely to produce false entity pairs, such as (u_5, v_3) in case (b).

Worse still, as pointed out in previous works [44, 59], existing representation learning methods for EA cannot fully capture the structural information (possibly due to their inner design mechanisms, or their incapability of dealing with scarce supervision signals). Under these settings, e.g., case (c), the distribution of entity embeddings in the low-dimensional space would become irregular, where the simple embedding matching algorithm DInf would fall short, i.e., producing incorrect entity pairs (u_3, v_1) and (u_5, v_1). As thus, in these practical cases, an effective embedding matching algorithm is crucial to inferring the correct matches. For instance, by exploiting the collective embedding matching algorithm that imposes the 1-to-1 alignment constraint, the correct matches, i.e., (u_3, v_3) and (u_5, v_5), are likely to be restored.

While the study on matching KGs in entity embedding spaces is rapidly progressing, there is no systematic survey or comparison of these solutions [44]. We do notice that there are several survey papers covering embedding-based EA

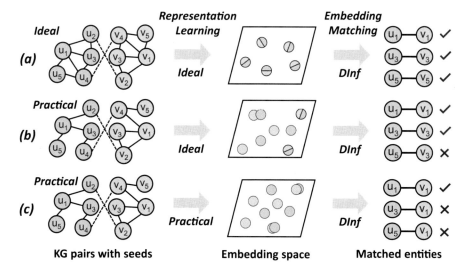

Fig. 4.1 Three cases of EA. Dashed lines between KGs denote the seed entity pairs. Entities with the same subscripts are equivalent. In the embedding space, the circles with two colors represent that the corresponding entities in the two KGs have the same embeddings

frameworks [44, 52, 57–59], whereas they all briefly introduce the embedding matching module (mostly only mentioning the DInf algorithm). In this chapter, we aim to fill in this gap by surveying current solutions for matching KGs in entity embedding spaces and providing a comprehensive empirical evaluation of these methods with the following features:

(1) **Systematic Survey and Fair Comparison** Albeit essential to the alignment performance, existing embedding matching strategies have yet not been compared directly. Instead, they are integrated with representation learning models and then evaluated and compared with each other (as a whole). This, however, cannot provide a fair comparison of the embedding matching strategies themselves, since the difference among them can be offset by other influential factors, such as the choices of representation learning models or input features. Therefore, in this chapter, we exclude irrelevant factors and provide a fair comparison of current matching algorithms for KGs in entity embedding spaces at both theoretical and empirical levels.

(2) **Comprehensive Evaluation and Detailed Discussion** To fully appreciate the effectiveness of embedding matching strategies, we conduct extensive experiments on a wide range of EA settings, i.e., with different representation learning models, with various input features, and on datasets at different scales. We also analyze the complexity of these algorithms and evaluate their efficiency/scalability under each experimental setting. Based on the empirical results, we discuss to reveal strengths and weaknesses.

(3) **New Experimental Settings and Insights** Through empirical evaluation and analysis, we discover that the current mainstream evaluation setting, i.e., 1-to-1 constrained EA, oversimplifies the real-life alignment scenarios. As thus, we identify two experimental settings that better reflect the challenges in practice, i.e., alignment with unmatchable entities, as well as a new setting of *non-1-to-1 alignment*. We compare the embedding matching algorithms under these challenging settings to provide further insights.

Contributions We make the following contributions:

- We systematically and comprehensively survey and compare state-of-the-art algorithms for *matching KGs in entity embedding spaces* (Sect. 4.3).
- We evaluate and compare the state-of-the-art embedding matching algorithms on a wide range of EA datasets and settings, as well as reveal their strengths and weaknesses. The codes of these algorithms are organized and integrated into an open-source library, EntMatcher, publicly available at https://github.com/DexterZeng/EntMatcher (Sect. 4.4).
- We identify experimental settings that better mirror real-life challenges and construct a new benchmark dataset, where deeper insights into the algorithms are obtained via empirical evaluations (Sect. 4.5).

- Based on our evaluation and analysis, we provide useful insights into the design trade-offs of existing works and suggest promising directions for the future development of matching KGs in entity embedding spaces (Sect. 4.6).

4.2 Preliminaries

In this section, we first present the task formulation of EA and its general framework. Next, we introduce the studies related to alignment inference and clarify the scope of this chapter. Finally, we present the key assumptions of embedding-based EA.

4.2.1 Task Formulation and Framework

Task Formulation A KG $\mathcal{G}$ is composed of triples $\{(s, p, o)\}$, where $s, o \in \mathcal{E}$ represent entities and $p \in \mathcal{P}$ denotes the predicate (relation). Given a source KG $\mathcal{G}_s$, a target KG $\mathcal{G}_t$, the task of EA is formulated as discovering new (equivalent) entity pairs $\mathcal{M} = \{(u, v) | u \in \mathcal{E}_s, v \in \mathcal{E}_t, u \Leftrightarrow v\}$ by using pre-annotated (seed) entity pairs $\mathcal{S}$ as anchors, where $\Leftrightarrow$ represents the equivalence between entities and $\mathcal{E}_s$ and $\mathcal{E}_t$ denote the entity sets in $\mathcal{G}_s$ and $\mathcal{G}_t$, respectively.

General Framework The pipeline of state-of-the-art embedding-based EA solutions can be divided into two stages, i.e., *representation learning* and *embedding matching*, as shown in Fig. 4.2. The general algorithm can be found in Algorithm 1.

The majority of studies on EA are devoted to the *representation learning* stage. They first utilize KG embedding techniques such as TransE [4] and GCN [20] to capture the KG structure information and generate entity structural representations. Next, based on the assumption that equivalent entities from different KGs possess similar neighboring KG structures (and in turn similar embeddings), they leverage the seed entity pairs as anchors and progressively project individual KG embeddings

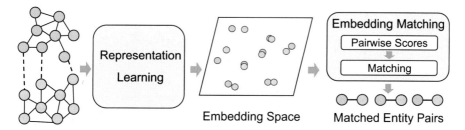

Fig. 4.2 The pipeline of embedding-based EA. Dashed lines denote the pre-annotated alignment links

---

Algorithm 1: General algorithm of embedding-based EA

   Input   : Source and target KGs: $\mathcal{G}_s, \mathcal{G}_t$; Seed pairs: S
   Output : Aligned entity pairs: $\mathcal{M}$
1 $E \leftarrow$ Representation_Learning($\mathcal{G}_s, \mathcal{G}_t, S$);
2 $\mathcal{M} \leftarrow$ Embedding_Matching($\mathcal{E}_s, \mathcal{E}_t, E$);
3 **return** $\mathcal{M}$;

---

into a unified space through training, resulting in the unified entity representations E.[2] There have already been several survey papers concentrating on representation learning approaches for EA, and we refer the interested readers to these works [2, 44, 57, 59].

Next, we introduce the *embedding matching* process –the focus of this chapter– as well as its related works.

4.2.2   Related Work and Scope

Matching KGs in Entity Embedding Spaces After obtaining the unified entity representations E where equivalent entities from different KGs are assumed to have similar embeddings, the *embedding matching* stage (also frequently referred to as *alignment inference* stage [44]) produces alignment results by comparing the embeddings of entities from different KGs. Concretely, it first calculates the pairwise scores between source and target entity embeddings according to a specific metric.[3] The pairwise scores are then organized into matrix form as S. Next, according to the pairwise scores, various matching algorithms are put forward to align entities. The most common algorithm is Greedy, described in Algorithm 2. It directly matches a source entity to the target entity that possesses the highest pairwise score according to S. Over the last few years, advanced solutions [12, 15, 28, 29, 44, 49, 53, 55, 56, 60] are devised to improve the embedding matching performance, and in this work, we focus on surveying and comparing these algorithms for matching KGs in entity embedding spaces.

Matching KGs in Symbolic Spaces Before the emergence of embedding-based EA, there have already been many conventional frameworks that match KGs in symbolic spaces [17, 41, 42]. While some are based on equivalence reasoning man-

---

[2] Indeed there are a few exceptions, which instead learn a mapping function between individual embedding spaces [44]. However, the subsequent steps still require mapping between spaces and operate on a "unified" one, e.g., target entity embeddings.

[3] Under certain metrics such as cosine similarity (resp., Euclidean distance), the larger (resp., smaller) the pairwise scores, the higher the probability that two entities are equivalent. In this work, w.l.o.g., we adopt the former expression and consider that higher pairwise scores are preferred.

Algorithm 2: Greedy($\mathcal{E}_s$, $\mathcal{E}_t$, S)

    Input   : Source and target entity sets: $\mathcal{E}_s$, $\mathcal{E}_t$; The similarity matrix of pairwise scores: S
    Output : Matched entity pairs: $\mathcal{M}$
1 for $u \in \mathcal{E}_s$ **do**
2 $v^* = \arg\max_{v \in \mathcal{E}_t} S(u, v)$;
3 $\mathcal{M} \leftarrow \mathcal{M} \cup \{(u, v^*)\}$;
4 return $\mathcal{M}$;

---

dated by OWL semantics [17], some leverage similarity computation to compare the symbolic features of entities [42]. However, these solutions are not comparable to algorithms for matching KGs in entity embedding spaces, as (1) they cover *both* the representation learning and embedding matching stages in embedding-based EA and (2) the required inputs are different from those of embedding matching algorithms. Thus, we do not include them in our experimental evaluation, while they have already been compared in the survey papers covering the overall embedding-based EA frameworks [44, 59].

The matching of relations (or ontology) between KGs has also been studied by prior symbolic works [41, 42]. Nevertheless, compared with entities, they are usually in smaller amounts, of various granularities [36], and under-explored in embedding-based approaches [51]. Hence, in this work, we exclude relevant studies on this topic and focus on the matching of *entities*.

The task of entity resolution (ER) [9, 16, 35], also known as entity matching, deduplication, or record linkage, can be regarded as the general case of EA [59]. It assumes that the input is relational data, and each data object usually has a large amount of textual information described in multiple attributes. Nevertheless, in this article, we focus on EA approaches, which strive to align *KGs* and mainly rely on *graph representation learning* techniques to model the KG structure and generate entity structural *embeddings* for alignment. Therefore, the discussion and comparison with ER solutions is beyond the scope of this work.

Matching Data Instances via Deep Learning Entity matching (EM) between databases has also been greatly advanced by utilizing pre-trained language models for expressive contextualization of database records [10, 33]. These deep learning (DL)-based EM solutions devise end-to-end neural models to learn to classify an entity pair into matching or non-matching and then feed the test entity pairs into the trained models to obtain classification results [5, 25, 33]. Nevertheless, this procedure is different from the focus of our study, as both of its training and testing stage involve representation learning and matching. Besides, these solutions are not suitable for matching KGs in entity embedding space, since (1) they require adequate labeled data to train the neural classification models, but the training data in EA is much less than the testing ones, which could result in the overfitting issue; (2) they would suffer from severe class imbalance in EA, where an entity and all of its nonequivalent entities in another KG would constitute many negative samples, while there is usually one positive sample for this entity; and (3) they depend on the attributive text information between data records for training, while EA underlines the use of KG structure, which could provide much less useful features for model training. In the experiment, we adapt DL-based EM models to tackle EA, and the results are not promising. This will be further discussed in Sect. 4.4.3.

Existing Surveys on EA There are several survey papers covering EA frameworks [44, 52, 57–59]. Some articles provide high-level discussion of embedding-based EA frameworks, experimentally evaluate and compare these works, and offer guidelines for potential practitioners [44, 58, 59]. Specifically, Zhao et al. propose a general EA framework to encompass existing works and then evaluate them under a wide range of settings. Nevertheless, they only briefly mention DInf and SMat in the embedding matching stage [59]. Sun et al. survey EA approaches and develop an open-source library to evaluate existing works. However, they merely introduce DInf, SMat, and Hun. and overlook the comparison among these algorithms. Besides, they point out that current approaches put in their main efforts in learning expressive embeddings to capture entity features but ignore the alignment inference (i.e., embedding matching) stage [44]. Zhang et al. empirically evaluate state-of-the-art embedding-based EA methods in an industrial context and particularly investigate the influence of the sizes and biases in seed mappings. They evaluate each method as a whole and do not mention the embedding matching process [58].

Two recent survey papers include the latest efforts on embedding-based EA and give more self-contained explanation on each technique. Zhang et al. provide a tutorial-type survey, while for embedding matching, they merely introduce the nearest neighbor search strategy, i.e., DInf [57]. Zeng et al. mainly introduce representation learning methods and their applications on EA, but they neglect the embedding matching stage [52].

In all, existing EA survey articles focus on the **representation learning** process and briefly introduce the embedding matching module (mostly only mentioning the DInf algorithm), while in this work, we systematically survey and empirically evaluate the algorithms designed for the **embedding matching** process in KG alignment and present comprehensive results and insightful discussions.

Scope of This Work This study aims to survey and empirically compare the algorithms for matching KGs in *entity embedding* spaces, i.e., various implementations of Embedding_Matching( ) in Algorithm 1, on a wide range of EA experimental settings.

4.2.3   Key Assumptions

Notably, existing embedding-based EA solutions have a **fundamental assumption**; that is, the equivalent entities in different KGs possess similar (ideally, isomorphic) neighboring structures. Under such an assumption, effective representation learning models would transform the structures of equivalent entities into similar entity embeddings. As thus, based on the entity embeddings, the embedding matching stage would assign higher (resp., lower) pairwise similarity scores to the equivalent (resp., nonequivalent) entity pairs and finally make accurate alignment decisions via the coordination according to pairwise scores.

Besides, current EA evaluation settings assume that the entities in different KGs conform to the 1-to-1 constraint. That is, each $u \in \mathcal{E}_s$ has *one and only one* equivalent entity $v \in \mathcal{E}_t$ and vice versa. However, we contend that this assumption is in fact impractical and provides detailed experiments and discussions in Sect. 4.5.2.

4.3 Alignment Inference Algorithms

In this section, we introduce the algorithms for alignment inference, i.e., Embedding_Matching( ) in Algorithm 1.

4.3.1 Overview

We first provide the overview and comparison of matching algorithms for KGs in entity embedding spaces in Table 4.1. As mentioned in Sect. 4.2, embedding matching comprises two stages—**pairwise score** computation and **matching**. The baseline approach DInf adopts existing similarity metrics to calculate the similarity between entity embeddings and generate the pairwise scores in the first stage, and then it leverages Greedy for matching. In pursuit of better alignment performance, more advanced embedding matching strategies are put forward. While some (i.e., CSLS, RInf, and Sink.) optimize the **pairwise score** computation process and produce more *accurate* pairwise scores, some (i.e., Hun., SMat, and RL) take into account the global alignment dynamics, rather than greedily pursue the local optimum for each entity, during the **matching** process, where more correct matches could be generated according to the coordination under the global constraint.

We further identify two notable characteristics of matching KGs in entity embedding spaces, i.e., whether the matching leverages the **1-to-1** constraint, and the **direction** of the matching. Regarding the former, Hun. and SMat explicitly exert the 1-to-1 constraint on the matching process. RL relaxes the strict 1-to-1 constraint by allowing non-1-to-1 matches. The greedy strategies, however, normally do not take into consideration this constraint, except for Sink., which implicitly implements the 1-to-1 constraint in a progressive manner when calculating the pairwise scores. As for the **direction** of matching, Greedy only considers a single direction at a time and overlooks the influence from the reverse direction. As thus, the resultant source-to-target alignment results are not necessarily equal to the target-to-source ones. By improving the pairwise score computation, CSLS, RInf, and Sink. are actually modeling and integrating the bidirectional alignments, whereas they still

Table 4.1 Overview and comparison of state-of-the-art algorithms for matching KGs in entity embedding spaces

Models	Pairwise scores	Matching	1-to-1	Direction	Time comp.	Space comp.
DInf. [24, 61]	Similarity metric	Greedy	✗	Unidirectional	$O(n^2)$	$O(n^2)$
CSLS [28, 44]	CSLS	Greedy	✗	Partially bidirectional	$O(n^2)$	$O(n^2)$
RInf. [53]	Preference modeling	Greedy	✗	Partially bidirectional	$O(n^2 \lg n)$	$O(n^2)$
Sink. [12, 15]	Sinkhorn operation	Greedy	Partially	Partially bidirectional	$O(ln^2)$	$O(n^2)$
Hun. [29, 49]	Similarity metric	Hungarian	✓	Bidirectional	$O(n^3)$	$O(n^2)$
SMat [55, 60]	Similarity metric	Gale-Shapley	✓	Bidirectional	$O(n^2 \lg n)$	$O(n^2)$
RL [56]	Similarity metric	Reinforcement learning	Partially	Unidirectional	/	$O(n^2)$

adopt Greedy to produce final results. For non-greedy methods, Hun. and SMat fully consider the bidirectional alignments and produce a matching agreed by both directions, while RL is unidirectional.

Next, we describe these methods in detail.[4]

4.3.2 Simple Embedding Matching

DInf is the most common implementation of Embedding_Matching( ), described in Algorithm 3. Assume both KGs contain n entities. The time and space complexity of DInf is $O(n^2)$.

Algorithm 3: DInf($\mathcal{E}_s, \mathcal{E}_t, E$)

    Input   : Source and target entity sets: $\mathcal{E}_s, \mathcal{E}_t$; Unified entity embeddings: E
    Output : Matched entity pairs: $\mathcal{M}$
1 Derive similarity matrix S based on E;
2 $\mathcal{M} \leftarrow$ Greedy($\mathcal{E}_s, \mathcal{E}_t, S$);
3 return $\mathcal{M}$;

4.3.3 CSLS Algorithm

The cross-domain similarity local scaling (CSLS) algorithm [23] is introduced to mitigate the hubness and isolation issues of entity embeddings in EA [44]. The hubness issue refers to the phenomenon where some entities (known as hubs) frequently appear as the top one most similar entities of other entities in the vector space, while the isolation issue means that there exist some outliers isolated from any point clusters. As thus, CSLS increases the similarity associated with isolated entity embeddings and conversely decreases the ones of vectors lying in dense areas [23]. Formally, the CSLS pairwise score between source entity u and target entity v is:

$$\text{CSLS}(u, v) = 2S(u, v) - \phi(u) - \phi(v)^\top, \tag{4.1}$$

where S is the similarity matrix derived from E using similarity metrics, $\phi(u) = \frac{1}{k}\sum_{v' \in \mathcal{N}_u} S(u, v')$ is the mean similarity score between the source entity u and its top-k most similar entities $\mathcal{N}_u$ in the target KG, and $\phi(v)$ is defined similarly. The mean similarity scores of all source and target entities are denoted in vector

---

[4] We omit the algorithmic description of the classical algorithms (e.g., Hungarian [21] and Gale-Shapley [40]) and the neural model (i.e., RL [32]) in the interest of space.

form as ϕ_s and ϕ_t, respectively. To generate the matched entity pairs, it further applies Greedy on the CSLS matrix (i.e., S_{CSLS}). Algorithm 4 describes the detailed procedure of CSLS.

Algorithm 4: CSLS($\mathcal{E}_s, \mathcal{E}_t, E, k$)

Input   : Source and target entity sets: $\mathcal{E}_s, \mathcal{E}_t$; Unified entity embeddings: E;
            Hyper-parameter: k
Output : Matched entity pairs: $\mathcal{M}$
1 Derive similarity matrix S based on E;
2 Calculate the mean values of top-k similarity scores of entities in $\mathcal{E}_s$ and $\mathcal{E}_t$, resulting in ϕ_s
   and ϕ_t, respectively;
3 $S_{\text{CSLS}} = 2S - \phi_s - \phi_t^\top$;
4 $\mathcal{M} \leftarrow$ Greedy($\mathcal{E}_s, \mathcal{E}_t, S_{\text{CSLS}}$);
5 **return** $\mathcal{M}$;

Complexity The time and space complexity is $O(n^2)$. Practically, it requires more time and space than DInf, as it needs to generate the additional CSLS matrix.

4.3.4 Reciprocal Embedding Matching

Zeng et al. [53] formulate EA task as the reciprocal recommendation process [38] and offer a reciprocal embedding matching strategy RInf to model and integrate the bidirectional preferences of entities when inferring the matching results. Formally, it defines the pairwise score of source entity u toward target entity v as:

$$p_{u,v} = S(u, v) - \max_{u' \in \mathcal{E}_s} S(v, u') + 1, \tag{4.2}$$

where S is the similarity matrix derived from E, $0 \leq p_{u,v} \leq 1$, and a larger $p_{u,v}$ denotes a higher degree of preference. As such, the matrix forms of the source-to-target and target-to-source preference scores are denoted as $P_{s,t}$ and $P_{t,s}$, respectively. Next, it converts the preference matrix P into the ranking matrix R and then averages the two ranking matrices, resulting in the reciprocal preference matrix $P_{s \leftrightarrow t}$ that encodes the bidirectional alignment information. Finally, it adopts Greedy to generate the matched entity pairs.

Complexity Algorithm 5 describes the detailed procedure of RInf. The time complexity is $O(n^2 \lg n)$ [53]. The space complexity is $O(n^2)$. Practically, it requires more space than DInf and CSLS, due to the computation of similarity, preference, and ranking matrices. Noteworthily, two variant methods, i.e., RInf-wr and RInf-pb, are proposed to reduce the memory and time consumption brought by the reciprocal modeling. More details can be found in [53].

Algorithm 5: RInf($\mathcal{E}_s$, $\mathcal{E}_t$, E)

Input : Source and target entity sets: $\mathcal{E}_s$, $\mathcal{E}_t$; Unified entity embeddings: E
Output : Matched entity pairs: $\mathcal{M}$
1 Derive similarity matrix S based on E;
2 **for** $u \in \mathcal{E}_s$ **do**
3   $\quad$ **for** $v \in \mathcal{E}_t$ **do**
4    $\quad\quad$ Calculate $p_{u,v}$ and $p_{v,u}$ (cf. Eq. (4.2));

5 Collect the preference scores, resulting in $P_{s,t}$ and $P_{t,s}$;
6 Convert $P_{s,t}$ and $P_{t,s}$ into $R_{s,t}$ and $R_{t,s}$, respectively;
7 $P_{s \leftrightarrow t} = (R_{s,t} + R_{t,s}^\top)/2$;
8 $\mathcal{M} \leftarrow$ Greedy($\mathcal{E}_s$, $\mathcal{E}_t$, $-P_{s \leftrightarrow t}$);
9 **return** $\mathcal{M}$;

4.3.5 Embedding Matching as Assignment

Some very recent studies [29, 49] propose to model the embedding matching process as the linear assignment problem. They first use similarity metrics to calculate pairwise similarity scores based on E. Then they adopt the Hungarian algorithm [21] to solve the task of assigning source entities to target entities according to the pairwise scores. The objective is to maximize the sum of the pairwise similarity scores of the final matched entity pairs while observing the 1-to-1 assignment constraint. In this work, we use the Hungarian algorithm implemented by Jonker and Volgenant [18] and denote it as Hun.($\mathcal{E}_s$, $\mathcal{E}_t$, E).

Besides, the Sinkhorn operation [31] (or Sink. for short) is also adopted to solve the assignment problem [12, 15, 29], which converts the similarity matrix S into a doubly stochastic matrix $S_{sinkhorn}$ that encodes the entity correspondence information. Specifically,

$$Sinkhorn^l(S) = \Gamma_c(\Gamma_r(Sinkhorn^{l-1}(S)));$$
$$S_{sinkhorn} = \lim_{l \to \infty} Sinkhorn^l(S), \qquad (4.3)$$

where $Sinkhorn^0(S) = \exp(S)$ and Γ_c and Γ_r refer to the column and row-wise normalization operators of a matrix. Since the number of iterations l is limited, the Sinkhorn operation can only obtain an approximate 1-to-1 assignment solution in practice [29]. Then $S_{sinkhorn}$ is forwarded to Greedy to obtain the alignment results.

Complexity For Hun., the time complexity is $O(n^3)$, and the space complexity is $O(n^2)$. Algorithm 5 describes the procedure of Sink.. The time complexity of Sink. is $O(ln^2)$ [29], and the space complexity is $O(n^2)$. In practice, both algorithms require more space than DInf, since they need to store the intermediate results.

Algorithm 6: Sink.$(\mathcal{E}_s, \mathcal{E}_t, E, l)$

Input   : Source and target entity sets: $\mathcal{E}_s, \mathcal{E}_t$; Unified entity embeddings: E;
            Hyper-parameter: l
Output : Matched entity pairs: $\mathcal{M}$
1 Derive similarity matrix S based on E;
2 $S_{sinkhorn} = Sinkhorn^l(S)$ (cf. Eq. (4.3));
3 $\mathcal{M} \leftarrow$ Greedy$(\mathcal{E}_s, \mathcal{E}_t, S_{sinkhorn})$;
4 **return** $\mathcal{M}$;

4.3.6 Stable Embedding Matching

In order to consider the interdependence among alignment decisions, the embedding matching process is formulated as the stable matching problem [13] by Zeng et al. [55] and Zhu et al. [60]. It is proved that for any two sets of members with the same size, each of whom provides a ranking of the members in the opposing set, there exists a bijection of the two sets such that no pair of two members from the opposite side would prefer to be matched to each other rather than their assigned partners [11]. Specifically, these works first produce the similarity matrix S based on E using similarity metrics. Next, they generate the rankings of members in the opposing set according to the pairwise similarity scores. Finally, they use the Gale-Shapley algorithm [40] to solve the stable matching problem. This procedure is denoted as SMat$(\mathcal{E}_s, \mathcal{E}_t, E)$.

Complexity SMat has time complexity of $O(n^2 \lg n)$ (since for each entity, the ranking of entities in the opposite side needs to be computed) and space complexity of $O(n^2)$.

4.3.7 RL-Based Embedding Matching

The embedding matching process is cast to the classic sequence decision problem by [56]. Given a sequence of source entities (and their embeddings), the goal of the sequence decision problem is to decide to which target entity each source entity aligns. It devises a reinforcement learning (RL)-based framework to learn to optimize the decision-making for all entities, rather than optimize every single decision separately. More details can be found in the original paper [56], and we use RL$(\mathcal{E}_s, \mathcal{E}_t, E)$ to denote it.

Complexity Note that it is difficult to deduce the time complexity for this neural RL model. Instead, we provide the empirical time costs in the experiments. The space complexity is $O(n^2)$.

4.4 Main Experiments

In this section, we compare the algorithms for matching KGs in entity embedding spaces on the mainstream EA evaluation setting (1-to-1 alignment).

4.4.1 EntMatcher: An Open-Source Library

To ensure comparability, we re-implemented all compared algorithms using Python under a unified framework and established an open-source library, EntMatcher.[5] The architecture of EntMatcher library is presented in the blue block of Fig. 4.3, which takes as input unified entity embeddings E and produces the matched entity pairs. It has the following three major features:

Loosely Coupled Design There are three independent modules in EntMatcher, and we have implemented the representative methods in each module. Users are free to combine the techniques in each module to develop new approaches, or to implement their new designs by following the templates in modules.

Reproduction of Existing Approaches To support our experimental study, we tried our best to re-implement all existing algorithms by using EntMatcher. For instance, the combination of cosine similarity, CSLS, and Greedy reproduces the CSLS algorithm in Sect. 4.3.3; and the combination of cosine similarity, None, and Hun. reproduces the Hun. algorithm in Sect. 4.3.5. The specific hyper-parameter settings are elaborated in Sect. 4.4.2.

Flexible Integration with Other Modules in EA EntMatcher is highly flexible, which can be directly called during the development of standalone EA approaches. Besides, users may also use EntMatcher as the backbone and call other modules. For instance, to conduct the experimental evaluations in this work, we implemented the representation learning and auxiliary information modules to generate the unified entity embeddings E, as shown in the white blocks of Fig. 4.3. More details are elaborated in the next subsection. Finally, EntMatcher is also compatible with existing open-source EA libraries (that mainly focus on representation learning) such as OpenEA[6] and EAkit.[7]

---

[5] The codes are publicly available at https://github.com/DexterZeng/EntMatcher.

[6] https://github.com/nju-websoft/OpenEA.

[7] https://github.com/THU-KEG/EAkit.

Fig. 4.3 Architecture of the
`EntMatcher` library and
additional modules required
by the experimental
evaluation

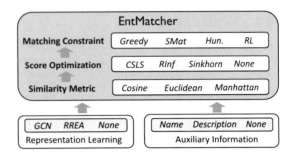

4.4.2   Experimental Settings

Current EA evaluation setting assumes that the entities in source and target KGs are
1-to-1 matched (cf. Sect. 4.2.3). Although this assumption simplifies the real-word
scenarios where some entities are unmatchable or some might be aligned to multiple
entities on the other side, it indeed reflects the core challenge of EA. Therefore, following existing literature, we mainly compare the embedding matching algorithms
under this setting and postpone the evaluation on the challenging real-life scenarios
to Sect. 4.5.

Datasets We used popular EA benchmarks for evaluation: **(1) DBP15K** , which
comprises three multilingual KG pairs extracted from DBpedia [1], English to
Chinese (DBP15K$_{ZH-EN}$), English to Japanese (DBP15K$_{JA-EN}$), and English to
French (DBP15K$_{FR-EN}$); **(2) SRPRS** , which is a sparser dataset that follows real-life
entity distribution, including two multilingual KG pairs extracted from DBpedia,
English to French (SRPRS$_{EN-FR}$) and English to German (SRPRS$_{EN-DE}$), and two
mono-lingual KG pairs, DBpedia to Wikidata [46] (SRPRS$_{DBP-WD}$) and DBpedia to
YAGO [43] (SRPRS$_{DBP-YG}$); and **(3) DWY100K** , which is a larger dataset consisting
of two mono-lingual KG pairs: DBpedia to Wikidata (D-W) and DBpedia to YAGO
(D-Y). The detailed statistics can be found in Table 4.2, where the numbers of
entities, relations, triples, gold links, and the average entity degree are reported.
Regarding the gold alignment links, we adopted 70% as test set, 20% for training,
and 10% for validation.

Hardware Configuration and Hyper-Parameter Setting Our experiments were
performed on a Linux server that is equipped with an Intel Core i7-4790 CPU
running at 3.6GHz, NVIDIA GeForce GTX TITAN X GPU, and 128 GB RAM.
We followed the configurations presented in the original papers of these algorithms
and tuned the hyper-parameters on the validation set. Specifically, for CSLS, we set
k to 1, except on the non-1-to-1 setting where we set it to 5. Similarly, regarding
RInf, we changed the maximum operation in Eq. (4.2) to top-k average operation
on the non-1-to-1 setting, where k is set to 5. As to Sink., we set l to 100. For RL,
we found the hyper-parameters in the original paper could already produce the best
results and directly adopted them. The rest of the approaches, i.e., DInf, Hun., and
SMat, do not contain hyper-parameters.

Table 4.2 Dataset statistics

	DBP15K			SRPRS				DWY100K		
	DBP15K$_{ZH-EN}$	DBP15K$_{JA-EN}$	DBP15K$_{FR-EN}$	SRPRS$_{EN-FR}$	SRPRS$_{EN-DE}$	SRPRS$_{DBP-WD}$	SRPRS$_{DBP-YG}$	D-W	D-Y	FB_DBP_MUL
#Entities	38,960	39,594	39,654	30,000	30,000	30,000	30,000	200,000	200,000	44,716
#Relations	3,024	2,452	2,111	398	342	397	253	550	333	2,070
#Triples	165,556	170,698	221,720	70,040	75,740	78,580	70,317	912,068	931,515	164,882
#Gold links	15,000	15,000	15,000	15,000	15,000	15,000	15,000	100,000	100,000	22,117
Avg. degree	4.2	4.3	5.6	2.3	2.5	2.6	2.3	4.6	4.7	3.7

Evaluation Metric We utilized *F1 score* as the evaluation metric, which is the harmonic mean between *precision* and *recall*, where the *precision* value is computed as the number of correct matches divided by the number of matches found by a method, and the *recall* value is computed as the number of correct matches found by a method divided by the number of gold matches. Note that *recall* is equivalent to the Hits@1 metric used in some previous works.

Representation Learning Models Since representation learning is not the focus of this work, we adopted two frequently used models, i.e., RREA [30] and GCN [47]. Concretely, GCN is one of the simplest models, which uses graph convolutional networks to learn the structural embeddings, while RREA is one of the best-performing solutions, which leverages relational reflection transformation to obtain relation-specific entity embeddings.

Auxiliary Information for Alignment Some works leverage the auxiliary information in KGs (e.g., entity attributes, descriptions, and pictures) to complement the KG structure. Specifically, these auxiliary information are first encoded into low-dimensional vectors and then fused with structural embeddings to provide more accurate entity representations for the subsequent embedding matching stage [44]. Although EA underlines the use of graph structure for alignment [59], for a more comprehensive evaluation, we examined the influence of auxiliary information on the matching results by following previous works and using entity name embeddings to facilitate alignment [34, 59]. We also combined these two channels of information with equal weights to generate the fused similarity matrix for matching.[8]

Similarity Metric After obtaining the unified entity representations E, a similarity metric is required to produce pairwise scores and generate the similarity matrix S. Frequent choices include the cosine similarity [6, 30, 45], the Euclidean distance [7, 24], and the Manhattan distance [48, 50]. In this work, we followed mainstream works and adopted the cosine similarity.

4.4.3 Main Results and Comparison

We first evaluate with only structural information and report the results in Table 4.3, where R- and G- refer to using RREA and GCN to generate the structural embeddings, respectively, and DBP and SRP denote DBP15K and SRPRS, respectively. Next, we supplement with name embeddings and report the results in Table 4.4, where N- and NR- refer to only using the name embeddings and fusing name embeddings with RREA structural representations, respectively. Note that, on

---

[8] Note that we only reported the results of fusing entity name embeddings with RREA structural embeddings. The results of combining entity name embeddings with GCN embeddings exhibited similar patterns and were omitted in the interest of space.

Table 4.3 The F1 scores of only using structural information

	R-DBP				R-SRP				
	DBP15K$_{ZH-EN}$	DBP15K$_{JA-EN}$	DBP15K$_{FR-EN}$	Imp.	SRPRS$_{EN-FR}$	SRPRS$_{EN-DE}$	SRPRS$_{DBP-WD}$	SRPRS$_{DBP-YG}$	Imp.
DInf	0.605	0.603	0.627		0.367	0.521	0.416	0.448	
CSLS	0.688	0.677	0.712	13.2%	0.406	0.550	0.465	0.481	8.6%
RInf	0.712	0.706	0.742	17.7%	0.412	0.560	0.477	0.486	10.4%
Sink.	**0.749**	0.740	**0.778**	23.5%	0.423	0.568	0.480	0.497	**12.3%**
Hun.	**0.749**	**0.744**	0.777	**23.7%**	0.418	0.563	0.475	0.495	11.4%
SMat	0.686	0.677	0.718	13.4%	0.398	0.551	0.453	0.471	6.9%
RL	0.675	0.670	0.716	12.3%	0.380	0.541	0.444	0.462	4.2%

	G-DBP				G-SRP				
	DBP15K$_{ZH-EN}$	DBP15K$_{JA-EN}$	DBP15K$_{FR-EN}$	Imp.	SRPRS$_{EN-FR}$	SRPRS$_{EN-DE}$	SRPRS$_{DBP-WD}$	SRPRS$_{DBP-YG}$	Imp.
DInf	0.291	0.295	0.286		0.170	0.322	0.202	0.253	
CSLS	0.375	0.390	0.377	31.0%	0.224	0.368	0.258	0.306	22.1%
RInf	0.400	0.423	0.423	42.9%	0.241	0.381	0.276	0.324	29.0%
Sink.	0.447	0.471	**0.484**	60.8%	0.248	0.387	**0.289**	**0.331**	**32.5%**
Hun.	**0.450**	**0.480**	**0.484**	**62.2%**	0.246	0.385	0.284	**0.331**	31.6%
SMat	0.382	0.413	0.388	35.7%	0.231	0.371	0.260	0.312	24.0%
RL	0.378	0.409	0.371	32.9%	0.213	0.361	0.245	0.288	16.9%

Table 4.4 The F1 scores of using auxiliary information

	N-DBP				N-SRP			NR-DBP				NR-SRP		
	DBP15K$_{ZH-EN}$	DBP15K$_{JA-EN}$	DBP15K$_{FR-EN}$	Imp.	SRPR$_{SEN-FR}$	SRPR$_{SEN-DE}$	Imp.	DBP15K$_{ZH-EN}$	DBP15K$_{JA-EN}$	DBP15K$_{FR-EN}$	Imp.	SRPR$_{SEN-FR}$	SRPR$_{SEN-DE}$	Imp.
DInf	0.735	0.780	0.744		0.815	0.831		0.819	0.862	0.846		0.865	0.893	
CSLS	0.754	0.802	0.761	2.6%	0.837	0.855	2.8%	0.858	0.896	0.880	4.2%	0.911	0.932	4.8%
RInf	0.751	0.802	0.761	2.4%	0.840	0.861	3.3%	0.861	0.899	0.887	4.7%	0.922	0.937	5.7%
Sink.	0.770	0.823	0.788	5.4%	0.853	0.878	5.2%	0.902	0.929	0.933	9.4%	0.940	0.954	7.7%
Hun.	**0.773**	**0.830**	**0.797**	**6.2%**	**0.864**	**0.887**	**6.4%**	**0.908**	**0.937**	**0.944**	**10.4%**	**0.949**	**0.956**	**8.4%**
SMat	0.768	0.818	0.778	4.6%	0.856	0.873	5.0%	0.879	0.912	0.906	6.7%	0.921	0.939	5.8%
RL	0.770	0.824	0.783	5.2%	0.851	0.866	4.3%	0.880	0.909	0.904	6.6%	0.917	0.936	5.4%

existing datasets, all the entities in the test set can be matched, and all the algorithms are devised to find a target entity for each test source entity. Hence, the number of matches found by a method equals to the number of gold matches, and consequently the precision value is equal to the recall value and the F1 score [56].

Overall Performance First, we do not delve into the embedding matching algorithms and directly analyze the general results. Specifically, using RREA to learn structural representations can bring better performance compared with using GCN, showcasing that representation learning strategies are crucial to the overall alignment performance. When introducing the entity name information, it observes that this auxiliary signal alone can already provide very accurate signal for alignment. This is because the equivalent entities in different KGs of current datasets share very similar or even identical names. After fusing the semantic and structural information, the alignment performance is further lifted, with most of the approaches hitting over 0.9 in terms of the F1 score.

Effectiveness Comparison of Embedding Matching Algorithms From the tables, it is evident that: (1) *Overall, Hun. and Sink. attain much better results than the other strategies.* Specifically, Hun. takes full account of the global matching constraints and strives to reach a globally optimal matching given the objective of maximizing the sum of pairwise similarity scores. Moreover, the 1-to-1 constraint it exerts aligns with present evaluation setting where the source and target entities are 1-to-1 matched. Sink., on the other hand, implicitly implements the 1-to-1 constraint during pairwise score computation and still adopts Greedy to produce final results, where there might exist non-1-to-1 matches; (2) *DInf attains the worst performance.* This is because it directly adopts the similarity scores that suffer from the hubness and isolation issues [44]. Besides, it leverages Greedy, which merely reaches the local optimum for each entity. (3) *The performance of RInf, CSLS, SMat, and RL are well matched.* RInf and CSLS improve upon DInf by mitigating the hubness issue and enhancing the quality of pairwise scores. SMat and RL, on the other hand, improve upon DInf by modeling the interactions among matching decisions for different entities.

Furthermore, we conduct a deeper analysis of these approaches and identify the following patterns:

Pattern 1. *If for source entities, their highest pairwise similarity scores are close, RInf and CSLS (resp., SMat and RL) would attain relatively better (resp., worse) performance.* Specifically, in Table 4.3 where RInf consistently (CSLS sometimes) attains superior results than SMat and RL, the average standard deviation (STD) values of the top five pairwise similarity scores of source entities (cf. Fig. 4.4) are very small, unveiling that the top scores are close and difficult to differentiate. In contrast, in Table 4.4 where SMat and RL outperform RInf and CSLS, the corresponding STD values are relatively large. This is because RInf and CSLS aim to make the scores more distinguishable, and hence they are more effective in cases where the top similarity scores are very close (i.e., low STD values). On the contrary, when the top similarity scores are already discriminating (e.g., Table 4.4), RInf and

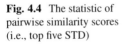

Fig. 4.4 The statistic of
pairwise similarity scores
(i.e., top five STD)

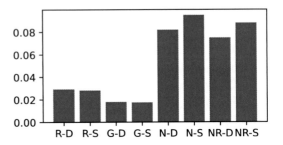

CSLS become less useful, while SMat and RL can still make improvements by using
the global constraints to enforce the deviation from local optimums.

Pattern 2. *On sparser datasets, the superiority of* Sink. *and* Hun. *over the
rest of the methods becomes less significant.* This is based on the observation
that on SRPRS, other matching algorithms (RInf in particular) attain much closer
performance to Sink. and Hun.. Such a pattern could be attributed to the fact that,
on sparser datasets, entities normally have fewer connections with others, i.e., lower
average entity degree (in Table 4.2), where representation learning strategies might
fail to fully capture the structural signals for alignment and the resultant pairwise
scores become less accurate. These inaccurate scores could mislead the matching
process and hence limit the effectiveness of the top-performing methods, i.e., Sink.
and Hun.. In other words, sparser KG structures are more likely to (partially) break
the fundamental assumption on KG structure similarity (cf. Sect. 4.2.3).

Efficiency Analysis We compare the time and space efficiency of these methods
on the medium-sized datasets in Fig. 4.5. Since the costs on KG pairs from the same
dataset are very similar, we report the *average* time and space costs under each
setting in the interest of space.

Specifically, it observes that: (1) The simple algorithm DInf is the most efficient
approach. (2) *Among the advanced approaches, CSLS is the most efficient one,*
closely following DInf. (3) *The efficiency of RInf and Hun. are equally matched.*
While Hun. consumes relatively less memory space than RInf, its time efficiency
is less stable and tends to run slower on datasets with less accurate pairwise scores.
(4) The space efficiency of Sink. is close to RInf and Hun., whereas it has much higher
time costs, which largely depends on the value of *l*. (5) *RL is the least time-efficient
approach, while SMat is the least space-efficient algorithm.* RL requires more time
on datasets with less accurate pairwise scores where its pre-processing module fails
to produce promising results [56]. The memory space consumption of SMat is high,
as it needs to store a large amount of intermediate matching results. In all, we can
conclude that *generally, advanced embedding matching algorithms require more
time and memory space, among which the methods incorporating global matching
constraints tend to be less efficient.*

Comparison with DL-Based EM Approaches We utilize the deepmatcher
Python package [33], which provides built-in neural networks and utilities that

Fig. 4.5 Efficiency
comparison. Shapes in blue
denote methods that improve
pairwise scores, while shapes
in black denote those exerting
global constraints (except for
DInf). (**a**) Time cost (in
seconds). (**b**) Memory space
cost (in GB)

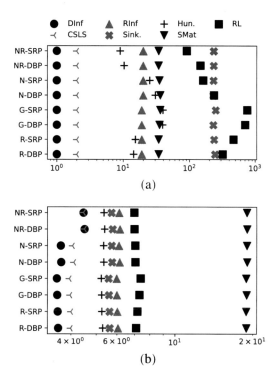

can train and apply state-of-the-art deep learning models for entity matching, to address EA. Specifically, we use the structural and name embeddings to replace the attributive text inputs in **deepmatcher**, respectively, and then train the neural model with labeled data. For each positive entity pair, we randomly sample ten negative ones. In the testing stage, for each source entity, we feed the entity pairs constituting it and all the target entities into the trained classifier and regard the entity pair with the highest predicted score as the result.

In the final results, only several entities are correctly aligned, showing that DL-based EM approaches cannot handle EA, which can be ascribed to the insufficient labeled data, imbalanced class distribution, and the lack of attributive text information, as discussed in Sect. 4.2.2.

4.4.4   Results on Large-Scale Datasets

Next, we provide the results on the relatively larger dataset, i.e., DWY100K, which can also reflect the scalability of these algorithms. The results are presented in

Table 4.5 The F1 scores on
DWY100K using GCN

	D-W	D-Y	Imp.	$\bar{T}$	Mem.
DInf	0.409	0.552		**4**	Yes
CSLS	0.510	0.650	20.7%	83	Yes
RInf	0.559	0.692	30.2%	1,102	No
RInf-wr	0.510	0.650	20.7%	28	Yes
RInf-pb	0.524	0.663	23.5%	289	Yes
Sink.	**0.618**	**0.739**	**41.2%**	9,405	No
Hun.	**0.618**	0.734	40.6%	3,607	No
SMat	/	/	/	/	/
RL	0.520	0.660	22.8%	995	Yes

Table 4.5.[9] The general pattern is similar to that on G-DBP (i.e., using GCN on DBP15K), where Sink. and Hun. obtain the best results, followed by RInf. The performance of CSLS and RL is close, outperforming DInf by over 20%.

We compare the efficiency of these algorithms in Table 4.5, where $\bar{T}$ refers to the average time cost and Mem. denotes whether the memory space required by the model can be covered by our experimental environment.[10] It observes that, given larger datasets, *most of the performant algorithms have poor efficiency and scalability* (e.g., RInf, Sink., and Hun.). Note that in [53], two variants of RInf, i.e., RInf-wr and RInf-pb, are proposed to improve its scalability at the cost of a small performance drop, which is empirically validated in Table 4.5. This also reveals that *more scalable matching algorithms for KGs in entity embedding spaces should be devised*.

4.4.5   Analysis and Insights

We provide further experiments and discussions in this subsection.

On Efficiency and Scalability The simple algorithm DInf is the most efficient and scalable one, as it merely involves the most basic computation and matching operations. CSLS is slightly less efficient than DInf due to the update of pairwise similarity scores. It also has good scalability. Although RInf adopts a similar idea to CSLS, it involves an additional ranking process, which brings much more time and memory consumption, making it less scalable. Sink. repeatedly conducts the normalization operation, and thus its time efficiency is mainly up to the l value. Its scalability is also limited by the memory space consumption since it needs to store intermediate results, as revealed in Table 4.5.

---

[9] We cannot provide the results of SMat, as it requires extremely large memory space and cannot work under our experimental environment.

[10] Note that for algorithms with memory space costs exceeding our experimental environment (except for SMat), there is additional swap area in the hard drive for them to finish the program (which usually takes much longer time).

Regarding the methods that exert global constraints, Hun. is efficient on medium-sized datasets, while it is not scalable due to the high time complexity and memory space consumption. SMat is space-inefficient even on the medium-sized datasets, making it not scalable. In comparison, RL has more stable time and space costs and can scale to large datasets, and the main influencing factor is the accuracy of pairwise scores. This is because RL has a pre-processing step that filters out confident matched entity pairs and excludes them from the time-consuming RL learning process [56]. More confident matched entity pairs would be filtered out if the pairwise scores are more accurate.

On Effectiveness of Improving Pairwise Score Computation   We compare and discuss the strategies for improving the pairwise score computation, i.e., CSLS, RInf, and Sink.

Both CSLS and RInf aim to mitigate the hubness and isolation issues in the raw pairwise scores (from different starting points). Particularly, we observe that, by setting k (in Eq. (4.1)) of CSLS to 1, *the difference between RInf and CSLS is reduced to the extra* ranking *process of RInf*, and the results in Table 4.3 and 4.4 validate that *this ranking process can consistently bring better performance*. This is because the ranking operation can amplify the difference among the scores and prevent such information from being lost after the bidirectional aggregation [53]. However, it is noteworthy that *the ranking process brings much more time and memory consumption*, as can be observed from the empirical results.

Then we analyze the influence of k value in CSLS. As shown in Fig. 4.6, a larger k leads to worse performance. This is because a larger k implies a smaller ϕ value in Eq. (4.1) (where the top-k highest scores are considered and averaged), and the resultant pairwise scores become less distinctive. This also validates the effectiveness of the design in RInf (cf. Eq. (4.2)), where only the *maximum* value is considered to compute the preference score. Nevertheless, in Sect. 4.5.2, we reveal that setting k to 1 is only useful in the 1-to-1 alignment setting.

As for Sink., it adopts an extreme approach to optimize the pairwise scores, which encourages each source (resp., target) entity to *have only one positive pairwise score* with a target (resp., source) entity and 0's with the rest of the target (resp., source) entities. Thus, *it is in fact progressively and implicitly implementing the 1-to-1 alignment constraint during the pairwise score computation process with*

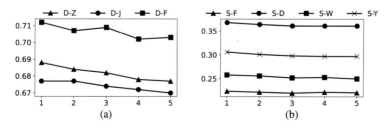

Fig. 4.6 F1 scores of CSLS with varying k value. (**a**) On R-DBP. (**b**) On G-SRP

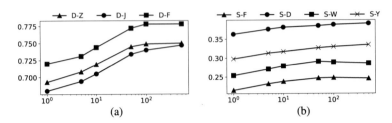

Fig. 4.7 F1 scores of Sink. with varying l value. (**a**) On R-DBP. (**b**) On G-SRP

the increase of l and is particularly useful in present 1-to-1 evaluation settings of EA. In Fig. 4.7, we further examine the influence of l in Eq. (4.3) on the alignment results of Sink., which meets our expectation that the larger the l value, the better the distribution of the resultant pairwise scores fits the 1-to-1 constraint, and thus the higher the alignment performance. Nevertheless, a larger l also implies longer processing time. Therefore, by tuning on the validation set, we set l to 100 to reach the balance between effectiveness and efficiency.

On Effectiveness of Exerting Global Constraints Next, we compare and discuss the methods that exert global constraints on the embedding matching process, i.e., Hun., SMat, and RL.

It is evident that *Hun. is the most performant approach, as it fits well with the present EA setting* and can secure an optimal solution toward maximizing the sum of pairwise scores. Specifically, the current EA setting has two notable assumptions (cf. Sect. 4.2.3). With these two assumptions, EA can be transformed into the linear assignment problem, which aims to maximize the sum of pairwise scores under the 1-to-1 constraint [29]. As thus, the algorithms for solving the linear assignment problem, e.g., Hun., can attain remarkably high performance on EA. However, these two assumptions do not necessarily hold on all occasions, which could influence the effectiveness of Hun.. For instance, as revealed in Pattern 2, on sparse datasets (e.g., SRPRS), the neighboring structures of some equivalent entities are likely to be different, where the effectiveness of Hun. is limited. In addition, the 1-to-1 alignment constraint is not necessarily true in practice, which will be discussed in Sect. 4.5.

In comparison, SMat merely aims to attain a stable matching, where the resultant entity pairing could be sub-optimal under present evaluation setting. RL, on the other hand, relaxes the 1-to-1 constraint and only deviates slightly from the greedy matching, and hence the results are not very promising.

Overall Comparison and Conclusion Finally, we compare the algorithms all together and draw the following conclusions *under the 1-to-1 alignment setting*: (1) The best performing methods are Hun. and Sink.. Nevertheless, they have low scalability. (2) CSLS and RInf achieve the best balance between effectiveness and efficiency. While CSLS is more efficient, RInf is more effective. (3) SMat and RL tend to attain better results when the accuracy of the pairwise scores is high. Nevertheless, they require relatively more time.

Table 4.6 F1 scores of combined embedding matching algorithms

	R–DBP	R–SRP	G–DBP	G–SRP	$\bar{T}$
CSLS	0.692	0.476	0.381	0.289	2
RInf	0.720	0.484	0.415	0.306	19
Hun.	0.757	0.488	0.471	0.312	27
CSLS +Hun.	0.757	0.488	0.471	0.312	20
RInf +Hun.	0.720	0.449	0.408	0.270	40
SMat	0.694	0.468	0.394	0.294	35
CSLS +SMat	0.706	0.476	0.411	0.304	33
RInf +SMat	0.722	0.476	0.411	0.300	52
RL	0.687	0.457	0.386	0.277	570
CSLS +RL	0.705	0.475	0.415	0.301	371
RInf +RL	0.716	0.472	0.399	0.294	206

Combining Embedding Matching Algorithms As described above, CSLS, RInf, and Sink. mainly improve the computation of pairwise scores, while Hun., SMat, and RL exert the global constraints during the matching process. Thus, by using the `EntMatcher` library, we aim to investigate whether the combination of these strategies would lead to better matching performance.

The results are reported in Table 4.6, where we can observe that: (1) Hun. is already effective given the raw pairwise scores, and using CSLS or RInf to improve the pairwise scores would not change and even bring down the performance (2) For SMat and RL, using CSLS or RInf to improve the raw pairwise scores would consistently lead to better results (3) Looking from the other side, while applying Hun., SMat, and RL upon CSLS improves its performance, such additional operations bring down the results of RInf This is because, by modeling entity preference and converting to rankings, RInf has already lost the information contained in the original pairwise scores, and exerting global constraints upon the reciprocal preference scores is no longer beneficial.

Hence, we can conclude that, generally speaking, *combining the algorithms (designed for different embedding matching stages) would lead to better alignment results*, except for the combination with RInf and Hun..

Alignment Results Analysis We further analyze the sets of matched entity pairs produced by the compared algorithms. Specifically, we examine the difference of the correct results found by different methods and report the pairwise *difference ratios* in the heat map of Fig. 4.8. The *difference ratio* is defined as the $|C - \mathcal{R}|/|C|$, where C and $\mathcal{R}$ denote the correct aligned entity pairs produced by the corresponding algorithms in the column and row, respectively.

From Fig. 4.8a, we can observe that: (1) The elements in the matrix (except those in secondary diagonal) are above 0, showing that these matching algorithms produce complementary correct matches (2) The results of Hun. and Sink., CSLS and RInf, and SMat and RL are similar; that is, the algorithms in each pair produce very similar correct matches (i.e., with low difference ratios and light colors) (3) The columns of

Fig. 4.8 Alignment results analysis on DBP15K$_{ZH-EN}$ of R-DBP. (**a**) Heat map of the difference ratios of correct matching results. (**b**) Proportions of correct results found by RInf, Hun., and SMat

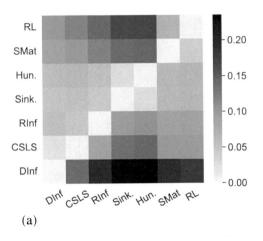

(a)

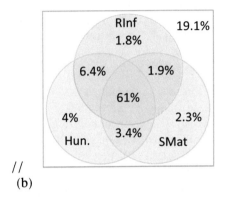

//
(b)

Hun., Sink., and RInf have relatively darker colors, revealing that they tend to discover the matches that other methods fail to detect

We further select three representative methods, i.e., RInf, Hun., and SMat, and provide a more detailed analysis in Fig. 4.8b. It is obvious that these approaches do produce complementary results, and *it calls for an ensemble framework that integrates the alignment results produced by different matching algorithms.*

4.5   New Evaluation Settings

In this section, we conduct experiments on settings that can better reflect real-life challenges.

4.5.1 Unmatchable Entities

Current EA literature largely overlooks the unmatchable issue, where a KG contains entities that the other KG does not contain. For instance, when aligning YAGO 4 and IMDB, only 1% of entities in YAGO 4 are film-related and possibly have equivalent entities in IMDB, while the other 99% of entities in YAGO 4 necessarily have no match in IMDB [59]. Hence, we aim to evaluate the embedding matching algorithms in terms of dealing with unmatchable entities.

Datasets and Evaluation Settings Following [54], we adapt the KG pairs in DBP15K to include unmatchable entities, resulting in DBP15K+. More specific construction procedure can be found in [54].

As for the evaluation metric, we follow the main experimental setting and adopt the *F1 score*. Unlike 1-to-1 alignment, there exist unmatchable entities in this adapted dataset, and the precision and recall values are not necessarily equivalent, since some methods would also align unmatchable entities.

Noteworthily, the original setting of SMat and Hun. requires that the numbers of entities on the two sides are equal. Thus, we add the dummy nodes on the side with fewer entities to restore such a setting and then apply SMat and Hun.. The corresponding results are reported in Table 4.7.

Alignment Results It reads that Hun. attains the best results, followed by SMat. The superior results are partially due to the addition of dummy nodes, which could mitigate the unmatchable issue to a certain degree. The results RInf and Sink. are close, outperforming CSLS and RL. DInf still achieves the worst performance.

Besides, by comparing the results on DBP15K+ and those on the original dataset DBP15K (cf. Table 4.3), we observe that: (1) After including the unmatchable entities, for all methods, the F1 scores drop. This is because most of current embedding matching algorithms are greedy, i.e., retrieving a target entity for each source entity (including the unmatchable ones), which leads to a very low precision. For the rest of the methods, e.g., Hun. and SMat, the unmatchable entities also mislead the matching process and thus affect the final results. (2) Unlike on DBP15K where the performance of Sink. and Hun. is close, on DBP15K+, Hun. largely outperforms Sink., as Hun. does not necessarily align a target entity to each source entity and has a higher precision. (3) *Overall, existing algorithms for matching KGs in entity embedding spaces lack the capability of dealing with unmatchable entities.*

4.5.2 Non-1-to-1 Alignment

Next, we study the setting where the source and target entities do not strictly conform to the 1-to-1 constraint, so as to better appreciate these matching algorithms for KGs in entity embedding spaces. Non-1-to-1 alignment is common in practice,

Table 4.7 F1 scores on DBP15K+

	GCN				RREA			
	DBP15K$_{\text{ZH-EN}}$	DBP15K$_{\text{JA-EN}}$	DBP15K$_{\text{FR-EN}}$	$\bar{T}$	DBP15K$_{\text{ZH-EN}}$	DBP15K$_{\text{JA-EN}}$	DBP15K$_{\text{FR-EN}}$	$\bar{T}$
DInf	0.241	0.240	0.234	**1**	0.501	0.491	0.513	**1**
CSLS	0.310	0.318	0.309	2	0.569	0.551	0.582	2
RInf	0.333	0.344	0.344	28	0.582	0.568	0.599	28
Sink.	0.329	0.337	0.343	336	0.571	0.553	0.584	331
Hun.	**0.397**	**0.407**	**0.408**	115	**0.712**	**0.706**	**0.750**	46
SMat	0.366	0.386	0.367	140	0.673	0.665	0.707	144
RL	0.307	0.311	0.297	1738	0.553	0.531	0.579	1264

Table 4.8 The results on non-1-to-1 alignment dataset

	GCN				RREA			
	P	R	F1	T	P	R	F1	T
DInf	0.074	0.051	0.061	**11**	0.167	0.114	0.136	**12**
CSLS	0.091	0.062	0.074	13	0.189	**0.130**	0.154	15
RInf	**0.093**	**0.064**	**0.076**	35	**0.190**	**0.130**	**0.155**	35
Sink.	0.083	0.057	0.068	286	0.180	0.124	0.147	278
Hun.	0.079	0.054	0.064	44	0.176	0.121	0.143	44
SMat	0.071	0.048	0.057	43	0.162	0.111	0.132	41
RL	0.066	0.045	0.054	1710	0.150	0.103	0.122	1440

especially when two KGs contain entities in different granularity, or one KG is noisy and involves duplicate entities. *To the best of our knowledge, we are among the first attempts to identify and investigate this issue.*

Dataset Construction Present EA benchmarks are constructed according to the 1-to-1 constraint. Thus, in this work, we establish a new dataset that involves non-1-to-1 alignment relationships. Specifically, we obtain the pre-annotated links[11] between Freebase [3] and DBpedia [1] and preserve the entities that are involved in 1-to-many, many-to-1, and many-to-many alignment relationships. Then, we retrieve the relational triples that contain these entities from respective KGs, which also introduces new entities.

Next, we detect the links among the newly added entities and add them into the alignment links. Finally, the resultant dataset, FB_DBP_MUL, contains 44,716 entities, 164,882 triples, and 22,117 gold links, among which 20,353 are non-1-to-1 links and 1,764 are 1-to-1 links.[12] The specific statistics are also presented in Table 4.2.

Evaluation Settings To keep the integrity of the links among entities, we sample the training, validation, and test sets from the gold links according to the principle that the links involving the same entity should not be distributed among different sets. The size of the final training, validation, and test sets is approximately 7:1:2. We compare the entity pairs produced by embedding matching algorithms against the gold test links and report the precision (P), recall (R), and F1 values.

Alignment Results It is evident from Table 4.8 that, compared with 1-to-1 alignment, the results change significantly on the new dataset. Specifically: (1) RInf and CSLS attain the best F1 scores, whereas the results are not very promising (e.g., with F1 score lower than 0.1 when using GCN). (2) Sink. and Hun. achieve much worse results compared with the performance on 1-to-1 alignment datasets. (3) The

---

[11] https://www.dbpedia.org/blog/dbpedia-is-now-interlinked-with-freebase-links-to-opencyc-updated/.

[12] FB_DBP_MUL is publicly available at https://github.com/DexterZeng/EntMatcher.

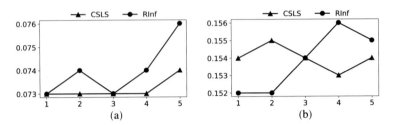

Fig. 4.9 F1 scores with varying k value on FB_DBP_MUL. (**a**) With GCN. (**b**) With RREA

results of SMat and RL are even inferior to those of the simple baseline DInf. The main reason accounting for these changes is that *the non-1-to-1 alignment links pose great challenges to existing embedding matching algorithms*. Specifically, for DInf, CSLS, RInf, Sink., and RL, they only align *one* target entity (that possesses the highest score) to a given source entity, but fail to discover other alignment links that also involve this source entity. For SMat and Hun., they impose the 1-to-1 constraint during the matching process, which falls short on the non-1-to-1 setting, thus leading to inferior results. Therefore, it calls for the study on embedding matching algorithms targeted at non-1-to-1 alignment.

Discussion of the k Value in CSLS and RInf In Fig. 4.9, we report the performance of CSLS and RInf given varying k values on FB_DBP_MUL. It shows that, generally, a larger k leads to better results. This is because, on the non-1-to-1 setting, an entity is likely to be matched to several entities on the other side, where it is more appropriate to consider the top-k values, rather than the sole maximum value, when refining the pairwise scores.

4.6 Summary and Future Direction

In this section, we summarize the observations and insights made from our evaluation and provide possible future research directions.

(1) The Investigation into Matching KGs in Embedding Spaces Has Not Yet Made Substantial Progress Although there are a few algorithms tailored for matching KGs in embedding spaces, e.g., CSLS, RInf, and RL, under the most popular EA evaluation setting (with 1-to-1 alignment constraint), they are outperformed by the classic general matching algorithms, i.e., Hun.. Hence, there is still much room for improving matching KGs in embedding spaces.

(2) No Existing Embedding Matching Algorithm Prevails Under All Experimental Settings The strategies designed to solve the linear assignment problem attain the best performance under the 1-to-1 setting, while they fall short on more practical and challenging scenarios since the new settings (e.g., non-1-to-1 alignment) no longer align with the conditions of these optimization algorithms. Similarly, although the methods for improving the computation of pairwise scores achieve superior results in the non-1-to-1 alignment scenario, they are outperformed by other solutions under the unmatchable setting. Therefore, each evaluation setting poses its own challenge to the embedding matching process, and currently there is no consistent winner.

(3) The Adaptation from General Matching Algorithms Requires Careful Design Among the embedding matching algorithms, Hun. and SMat are general matching algorithms that have been applied to many other related tasks. Although directly adopting these general strategies to tackle EA is simple and effective, they might well fall short in some scenarios, as the alignment on KGs possesses its own challenges, e.g., the matching is not necessarily 1-to-1 constrained, or the pairwise scores are inaccurate. Thus, it is suggested to take full account of the characteristics of the alignment settings when adapting other general matching algorithms to cope with matching KGs in entity embedding spaces.

(4) The Scalability and Efficiency Should Be Brought to the Attention Existing advanced embedding matching algorithms have poor scalability, due to the additional resource-consuming operations that contribute to the alignment performance, such as the ranking process in RInf and the 1-to-1 constraint exerted by Hun. and SMat. Besides, the space efficiency is also a critical issue. As shown in Sect. 4.4.4, most of the approaches have rather high memory costs given large-scale datasets. Therefore, considering that in practice there are much more entities, the scalability and efficiency issues should be considered during the algorithm design.

(5) The Practical Evaluation Settings Are Worth Further Investigation Under the unmatchable and non-1-to-1 alignment settings, the performance of existing algorithms is not promising. A possible future direction is to introduce the notion of probability and leverage the probabilistic reasoning frameworks [19, 39], which have higher flexibility, to produce the alignment results.

4.7 Conclusion

This paper conducts a comprehensive survey and evaluation of matching algorithms for KGs in entity embedding spaces. We evaluate seven state-of-the-art strategies in terms of effectiveness and efficiency on a wide range of datasets, including two experimental settings that better mirror real-life challenges. We identify the strengths and weaknesses of these algorithms under different settings. We hope the experimental results would be valuable for researchers and practitioners to put forward more effective and scalable embedding matching algorithms.

References

1. S. Auer, C. Bizer, G. Kobilarov, J. Lehmann, R. Cyganiak, and Z. G. Ives. Dbpedia: A nucleus for a web of open data. In *ISWC*, pages 722–735, 2007.
2. M. Berrendorf, E. Faerman, V. Melnychuk, V. Tresp, and T. Seidl. Knowledge graph entity alignment with graph convolutional networks: Lessons learned. In *ECIR*, volume 12036, pages 3–11, 2020.
3. K. D. Bollacker, C. Evans, P. Paritosh, T. Sturge, and J. Taylor. Freebase: a collaboratively created graph database for structuring human knowledge. In *SIGMOD*, pages 1247–1250, 2008.
4. A. Bordes, N. Usunier, A. García-Durán, J. Weston, and O. Yakhnenko. Translating embeddings for modeling multi-relational data. In *NIPS*, pages 2787–2795, 2013.
5. U. Brunner and K. Stockinger. Entity matching with transformer architectures - A step forward in data integration. In *Proceedings of the 23rd International Conference on Extending Database Technology, EDBT 2020, Copenhagen, Denmark, March 30 - April 02, 2020*, pages 463–473. OpenProceedings.org, 2020.
6. Y. Cao, Z. Liu, C. Li, Z. Liu, J. Li, and T. Chua. Multi-channel graph neural network for entity alignment. In *ACL*, pages 1452–1461, 2019.
7. M. Chen, Y. Tian, K. Chang, S. Skiena, and C. Zaniolo. Co-training embeddings of knowledge graphs and entity descriptions for cross-lingual entity alignment. In *IJCAI*, pages 3998–4004, 2018.
8. M. Chen, Y. Tian, M. Yang, and C. Zaniolo. Multilingual knowledge graph embeddings for cross-lingual knowledge alignment. In *IJCAI*, pages 1511–1517, 2017.
9. V. Christophides, V. Efthymiou, T. Palpanas, G. Papadakis, and K. Stefanidis. An overview of end-to-end entity resolution for big data. *ACM Comput. Surv.*, 53(6):127:1–127:42, 2021.
10. A. Doan, P. Konda, P. S. G. C., Y. Govind, D. Paulsen, K. Chandrasekhar, P. Martinkus, and M. Christie. Magellan: toward building ecosystems of entity matching solutions. *Commun. ACM*, 63(8):83–91, 2020.
11. J. Doerner, D. Evans, and A. Shelat. Secure stable matching at scale. In *SIGSAC Conference*, pages 1602–1613, 2016.
12. M. Fey, J. E. Lenssen, C. Morris, J. Masci, and N. M. Kriege. Deep graph matching consensus. In *ICLR*, 2020.
13. D. Gale and L. S. Shapley. College admissions and the stability of marriage. *The American Mathematical Monthly*, 69(1):9–15, 1962.
14. C. Ge, X. Liu, L. Chen, B. Zheng, and Y. Gao. Largeea: Aligning entities for large-scale knowledge graphs. *CoRR*, abs/2108.05211, 2021.
15. C. Ge, X. Liu, L. Chen, B. Zheng, and Y. Gao. Make it easy: An effective end-to-end entity alignment framework. In *SIGIR*, pages 777–786, 2021.
16. C. Ge, P. Wang, L. Chen, X. Liu, B. Zheng, and Y. Gao. Collaborem: A self-supervised entity matching framework using multi-features collaboration. *IEEE Transactions on Knowledge and Data Engineering*, pages 1–1, 2021.
17. E. Jiménez-Ruiz and B. C. Grau. Logmap: Logic-based and scalable ontology matching. In *ISWC*, pages 273–288. Springer, 2011.
18. R. Jonker and A. Volgenant. A shortest augmenting path algorithm for dense and sparse linear assignment problems. *Computing*, 38(4):325–340, 1987.
19. A. Kimmig, A. Memory, R. J. Miller, and L. Getoor. A collective, probabilistic approach to schema mapping. In *ICDE*, pages 921–932, 2017.
20. T. N. Kipf and M. Welling. Semi-supervised classification with graph convolutional networks. In *ICLR*. OpenReview.net, 2017.
21. H. W. Kuhn. The hungarian method for the assignment problem. *Naval research logistics quarterly*, 2(1–2):83–97, 1955.

22. S. Lacoste-Julien, K. Palla, A. Davies, G. Kasneci, T. Graepel, and Z. Ghahramani. Sigma: simple greedy matching for aligning large knowledge bases. In *SIGKDD*, pages 572–580. ACM, 2013.
23. G. Lample, A. Conneau, M. Ranzato, L. Denoyer, and H. Jégou. Word translation without parallel data. In *ICLR*, 2018.
24. C. Li, Y. Cao, L. Hou, J. Shi, J. Li, and T. Chua. Semi-supervised entity alignment via joint knowledge embedding model and cross-graph model. In *EMNLP*, pages 2723–2732, 2019.
25. Y. Li, J. Li, Y. Suhara, A. Doan, and W. Tan. Deep entity matching with pre-trained language models. *Proc. VLDB Endow.*, 14(1):50–60, 2020.
26. X. Lin, H. Yang, J. Wu, C. Zhou, and B. Wang. Guiding cross-lingual entity alignment via adversarial knowledge embedding. In *ICDM*, pages 429–438, 2019.
27. B. Liu, H. Scells, G. Zuccon, W. Hua, and G. Zhao. Activeea: Active learning for neural entity alignment. In *EMNLP*, pages 3364–3374, 2021.
28. X. Mao, W. Wang, Y. Wu, and M. Lan. Boosting the speed of entity alignment 10 ×: Dual attention matching network with normalized hard sample mining. In *WWW*, pages 821–832, 2021.
29. X. Mao, W. Wang, Y. Wu, and M. Lan. From alignment to assignment: Frustratingly simple unsupervised entity alignment. In *EMNLP*, pages 2843–2853, 2021.
30. X. Mao, W. Wang, H. Xu, Y. Wu, and M. Lan. Relational reflection entity alignment. In *CIKM*, pages 1095–1104, 2020.
31. G. E. Mena, D. Belanger, S. W. Linderman, and J. Snoek. Learning latent permutations with gumbel-sinkhorn networks. In *ICLR*, 2018.
32. V. Mnih, A. P. Badia, M. Mirza, A. Graves, T. P. Lillicrap, T. Harley, D. Silver, and K. Kavukcuoglu. Asynchronous methods for deep reinforcement learning. In *ICML*, volume 48, pages 1928–1937, 2016.
33. S. Mudgal, H. Li, T. Rekatsinas, A. Doan, Y. Park, G. Krishnan, R. Deep, E. Arcaute, and V. Raghavendra. Deep learning for entity matching: A design space exploration. In *Proceedings of the 2018 International Conference on Management of Data, SIGMOD Conference 2018, Houston, TX, USA, June 10–15, 2018*, pages 19–34. ACM, 2018.
34. T. T. Nguyen, T. T. Huynh, H. Yin, V. V. Tong, D. Sakong, B. Zheng, and Q. V. H. Nguyen. Entity alignment for knowledge graphs with multi-order convolutional networks. *IEEE Transactions on Knowledge and Data Engineering*, pages 1–1, 2020.
35. G. Papadakis, D. Skoutas, E. Thanos, and T. Palpanas. Blocking and filtering techniques for entity resolution: A survey. *ACM Comput. Surv.*, 53(2):31:1–31:42, 2020.
36. H. Paulheim. Knowledge graph refinement: A survey of approaches and evaluation methods. *Semantic Web*, 8(3):489–508, 2017.
37. S. Pei, L. Yu, and X. Zhang. Improving cross-lingual entity alignment via optimal transport. In S. Kraus, editor, *IJCAI*, pages 3231–3237, 2019.
38. L. A. S. Pizzato, T. Rej, T. Chung, I. Koprinska, and J. Kay. RECON: a reciprocal recommender for online dating. In *RecSys*, pages 207–214. ACM, 2010.
39. J. Pujara, H. Miao, L. Getoor, and W. W. Cohen. Large-scale knowledge graph identification using PSL. In *AAAI*, 2013.
40. A. E. Roth. Deferred acceptance algorithms: history, theory, practice, and open questions. *Int. J. Game Theory*, 36(3–4):537–569, 2008.
41. P. Shvaiko and J. Euzenat. Ontology matching: State of the art and future challenges. *IEEE Trans. Knowl. Data Eng.*, 25(1):158–176, 2013.
42. F. M. Suchanek, S. Abiteboul, and P. Senellart. PARIS: probabilistic alignment of relations, instances, and schema. *Proc. VLDB Endow.*, 5(3):157–168, 2011.
43. F. M. Suchanek, G. Kasneci, and G. Weikum. Yago: a core of semantic knowledge. In *WWW*, pages 697–706, 2007.
44. Z. Sun, Q. Zhang, W. Hu, C. Wang, M. Chen, F. Akrami, and C. Li. A benchmarking study of embedding-based entity alignment for knowledge graphs. *Proc. VLDB Endow.*, 13(11):2326–2340, 2020.

45. B. D. Trisedya, J. Qi, and R. Zhang. Entity alignment between knowledge graphs using attribute embeddings. In *AAAI*, pages 297–304, 2019.
46. D. Vrandecic and M. Krötzsch. Wikidata: a free collaborative knowledgebase. *Commun. ACM*, 57(10):78–85, 2014.
47. Z. Wang, Q. Lv, X. Lan, and Y. Zhang. Cross-lingual knowledge graph alignment via graph convolutional networks. In *EMNLP*, pages 349–357, 2018.
48. Y. Wu, X. Liu, Y. Feng, Z. Wang, R. Yan, and D. Zhao. Relation-aware entity alignment for heterogeneous knowledge graphs. In *IJCAI*, pages 5278–5284, 2019.
49. K. Xu, L. Song, Y. Feng, Y. Song, and D. Yu. Coordinated reasoning for cross-lingual knowledge graph alignment. In *AAAI*, pages 9354–9361, 2020.
50. H. Yang, Y. Zou, P. Shi, W. Lu, J. Lin, and X. Sun. Aligning cross-lingual entities with multi-aspect information. In *EMNLP*, pages 4430–4440, 2019.
51. J. Yang, D. Wang, W. Zhou, W. Qian, X. Wang, J. Han, and S. Hu. Entity and relation matching consensus for entity alignment. In *CIKM*, pages 2331–2341. ACM, 2021.
52. K. Zeng, C. Li, L. Hou, J. Li, and L. Feng. A comprehensive survey of entity alignment for knowledge graphs. *AI Open*, 2:1–13, 2021.
53. W. Zeng, X. Zhao, X. Li, J. Tang, and W. Wang. On entity alignment at scale. *VLDB J.*, page in press, 2021.
54. W. Zeng, X. Zhao, J. Tang, X. Li, M. Luo, and Q. Zheng. Towards entity alignment in the open world: An unsupervised approach. In *DASFAA 2021*, volume 12681, pages 272–289, 2021.
55. W. Zeng, X. Zhao, J. Tang, and X. Lin. Collective entity alignment via adaptive features. In *ICDE*, pages 1870–1873. IEEE, 2020.
56. W. Zeng, X. Zhao, J. Tang, X. Lin, and P. Groth. Reinforcement learning-based collective entity alignment with adaptive features. *ACM Trans. Inf. Syst.*, 39(3):26:1–26:31, 2021.
57. R. Zhang, B. D. Trisedya, M. Li, Y. Jiang, and J. Qi. A comprehensive survey on knowledge graph entity alignment via representation learning. *CoRR*, abs/2103.15059, 2021.
58. Z. Zhang, H. Liu, J. Chen, X. Chen, B. Liu, Y. Xiang, and Y. Zheng. An industry evaluation of embedding-based entity alignment. In *Proceedings of the 28th International Conference on Computational Linguistics, COLING 2020 - Industry Track, Online, December 12, 2020*, pages 179–189. International Committee on Computational Linguistics, 2020.
59. X. Zhao, W. Zeng, J. Tang, W. Wang, and F. Suchanek. An experimental study of state-of-the-art entity alignment approaches. *IEEE TKDE*, pages 1–1, 2020.
60. R. Zhu, M. Ma, and P. Wang. RAGA: relation-aware graph attention networks for global entity alignment. In *PAKDD*, volume 12712, pages 501–513, 2021.
61. Y. Zhu, H. Liu, Z. Wu, and Y. Du. Relation-aware neighborhood matching model for entity alignment. In *AAAI*, pages 4749–4756, 2021.

Part III
Novel Approaches

Chapter 5
Large-Scale Entity Alignment

Abstract In this chapter, we focus on the concept of entity alignment at scale and present a new method for addressing this task. The proposed solution is capable of handling vast amounts of knowledge graph pairs and delivering high-quality alignment outcomes. First, to manage large-scale KG pairs, we develop a set of seed-oriented graph partition strategies that divide them into smaller subgraph pairs. Next, within each subgraph pair, we employ existing methods to learn unified entity representations and introduce a novel reciprocal alignment inference strategy to model bidirectional alignment interactions, which can lead to more accurate outcomes. To further enhance the scalability of reciprocal alignment inference, we propose two variant strategies that can significantly reduce memory and time costs, albeit at the expense of slightly reduced effectiveness. Our solution is versatile and can be applied to existing representation learning-based EA models to enhance their ability to handle large-scale KG pairs. We also create a new EA dataset that comprises millions of entities and conduct comprehensive experiments to verify the efficiency of our proposed model. Furthermore, we compare our proposed model against state-of-the-art baselines on popular EA datasets, and our extensive experiments demonstrate its effectiveness and superiority.

5.1 Introduction

Figure 5.1 describes a toy example of EA. Typically, state-of-the-art EA solutions follow a two-stage working pipeline, which can be broadly divided into two main stages—*representation learning* and *alignment inference*. Most of the current works [4, 6, 33, 35] are dedicated to the former, which leverage various KG embedding models, e.g., TransE [2] and graph convolutional network (GCN) [15], for learning the representations of entities. By using seed entity pairs as reference points, the entity embeddings of different KGs are projected onto a common embedding space. This allows for the measurement of similarity or distance[1]

---

[1] In the rest of the paper, we may use "distance" and "similarity" with obvious adaptation.

© The Author(s) 2023
X. Zhao et al., *Entity Alignment*, Big Data Management,
https://doi.org/10.1007/978-981-99-4250-3_5

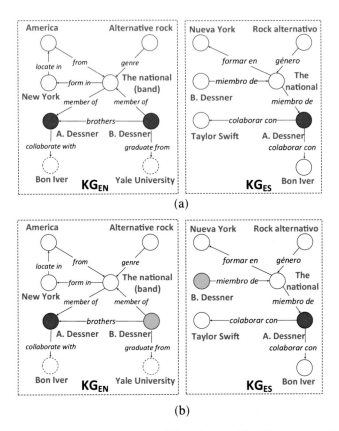

Fig. 5.1 An example of EA. There is an English and a Spanish KG concerning the band *The national* in the figures. The aim of EA is to find equivalent entities in these KGs using the KG structure, e.g., [A. Dessner]$_{en}$ and [A. Dessner]$_{es}$. The left denotes the alignment results generated by current EA solutions that perform direct alignment inference based on structural similarity, where both [A. Dessner]$_{en}$ and [B. Dessner]$_{en}$ are aligned to [A. Dessner]$_{es}$. In comparison, the right denotes the results generated by our proposed reciprocal alignment inference, where [A. Dessner]$_{en}$ is aligned to [A. Dessner]$_{es}$, while [B. Dessner]$_{en}$ is matched with [B. Dessner]$_{es}$. (**a**) Results of direct alignment inference using structural similarity. (**b**) Results of reciprocal alignment inference using entity preference

between entities from different KGs by assessing the similarity or distance between data points in the unified embedding space. Once the entities have been projected onto the unified embedding space, the *alignment inference* stage involves predicting the alignment results using these embeddings. When given an entity from the source KG, most state-of-the-art solutions adopt the *direct* alignment inference strategy. This strategy involves ranking the entities in the target KG based on a specific similarity measure between entity embeddings. The top-ranked target entity is then considered a match for the source entity.

Despite the improvements made by current techniques in boosting the precision of EA, these sophisticated models typically involve a substantial number of parameters and demand significant computational resources. Therefore, scalability is compromised in achieving the improvement, and these approaches are not suitable for handling practical large KGs. For instance, it is reported in [50] that, on the DWY100K dataset with 200,000 entities [33], the time cost for most state-of-the-art solutions is over 20,000 seconds, and some approaches [39, 52] even cannot produce the alignment results. Thus, the existence of real-life KGs that consist of tens of millions of entities creates a significant obstacle for current EA solutions, necessitating research on *large-scale entity alignment*. The investigation of large-scale EA aligns with the current trend of responsible design, development, use, and oversight of automated decision systems in the data management community [29].

Drawing inspiration from traditional graph partitioning strategies [12, 14], a feasible technique is to divide large KG pairs into several smaller subgraph pairs and then perform entity alignment on them. However, partitioning KG pairs for alignment is a challenging task that must achieve two objectives: (1) preserving the original structure of the KG as much as possible and (2) ensuring that the partition results of the source and target KGs match, meaning that equivalent entities in the source and target KGs are placed in the same subgraph pair. Although the first objective can be accomplished by modifying classical graph partitioning techniques such as METIS [13], the second objective is specific to the alignment task.

To achieve the second objective, we can use the seed entity pairs to guide the partition process. Seed entity pairs are pre-labeled entity pairs in which the entities are equivalent and are used to link two individual KGs. Ideally, if we can preserve the seed entity pairs during the partition and distribute them among the smaller subgraph pairs, the remaining (unknown) equivalent entities would have a greater likelihood of being placed in the same subgraph pair using these seed entity pairs as references, as equivalent entities usually have similar neighboring structures. Following this idea, there is a preliminary approach METIS-CPS (shortened as CPS) proposed by a concurrent work [10]. The proposed approach first partitions one KG into subgraphs. Then, based on the distribution of seed entities, it assigns appropriate weights to the edges in the other KG, and the partition is performed on this KG. However, it can be challenging for methods of this type (referred to as *unidirectional* partition strategies) to achieve the first objective because the partitioning of the second KG is limited by the requirement to maintain the seed links, which may compromise the structure of the KG to some extent.

To address this issue, this chapter proposes the Seed-oriented Bidirectional graph Partition framework, SBP, which aims to satisfy both objectives by conducting bidirectional partitions and aggregating the partition results from the source-to-target and target-to-source directions. The motivation behind this approach is that the subgraphs generated from partitioning the first KG tend to have more complete structures, while the subgraphs generated from partitioning the second KG mainly retain alignment signals. By performing bidirectional partitions and combining the subgraphs, the resulting subgraphs in each KG can have both complete structures and larger numbers of seed entities pointing to the subgraphs in the opposite KG, which can lead to more precise alignment results. Note that SBP can be used with various *unidirectional* partitioning strategies. Additionally, an iterative variant of SBP, I-SBP, is proposed to improve partition performance by incorporating confident alignment results from previous rounds into the seed entity pairs.

During the partition process, the accuracy of alignment results may be compromised because equivalent entities could be placed in different subgraph pairs, and the original KG structure information may also be lost to some extent. To improve alignment performance, we propose to enhance the alignment inference stage, which has received little attention in previous work. Specifically, we introduce a reciprocal alignment inference strategy. The idea of reciprocal modeling of the alignment process is motivated by the fact that the commonly used direct alignment inference approach (1) considers an entity's preference toward entities on the other side via a similarity score, but neglects other influential factors, and (2) fails to integrate bidirectional preference scores or capture the mutual preferences of entities when making alignment decisions. Such an alignment inference strategy tends to produce many inaccurate results, as illustrated in the following example.

Example As shown in Fig. 5.1a, using the structural information, the direct alignment inference strategy would align both [A. Dessner]$_{en}$ and [B. Dessner]$_{en}$ to [A. Dessner]$_{es}$, since [A. Dessner]$_{es}$ is the entity that has the most similar structural information with them (connected to three entities, including [The national]$_{en/es}$).

However, this direct inference approach overlooks the fact that entities' preferences are not solely determined by the similarity score but also by the impact of alignment in the reverse direction. For instance, it is evident that [A. Dessner]$_{es}$ has higher similarity with [A. Dessner]$_{en}$ than [B. Dessner]$_{en}$ since it shares more neighboring information with [A. Dessner]$_{en}$. Under this circumstance, [B. Dessner]$_{en}$ will lower its preference toward [A. Dessner]$_{es}$, since in its view, although [A. Dessner]$_{es}$ is its most preferred candidate in terms of similarity, they are less likely to form a match because [A. Dessner]$_{es}$ has a higher similarity with [A. Dessner]$_{en}$.

Therefore, by modeling and aggregating the bidirectional preferences as depicted in Fig. 5.1b, we could avoid matching [B. Dessner]$_{en}$ with [A. Dessner]$_{es}$ and possibly help identify its correct equivalent entity [B. Dessner]$_{es}$.

Specifically, we propose to model the entity alignment task as a reciprocal recommendation process [18, 27], which takes effect at two levels: (1) *Entity preference modeling.* It first incorporates the influence of the alignment in the reverse direction into an entity's preference, so as to generate more accurate preference scores. (2) *Bidirectional preference integration.* It integrates bidirectional preferences to generate a reciprocal preference matrix that encodes the mutual preferences of entities on both sides. Experimental results have shown that the two-level reciprocal modeling approach achieves superior results compared to direct inference (to be detailed in Sect. 5.7).

We further notice that while the reciprocal inference approach achieves superior alignment performance, it also consumes more memory space and time compared to direct alignment inference. Therefore, to improve the efficiency, we propose two variants: no-ranking aggregation and progressive blocking, which approximate the reciprocal alignment inference. While the former removes the time- and resource-consuming ranking process during the preference aggregation process, the latter divides the entities into multiple blocks and performs alignment within each block. These variant strategies can significantly reduce the memory and time costs associated with the reciprocal alignment inference, albeit at the cost of a slight decrease in effectiveness.

The proposed techniques form a novel and scalable solution for Large-scale entIty alignMEnt, namely, LIME. Notably, LIME is model-agnostic and can be used with any entity representation learning models. In this work, we evaluate using the commonly used GCN model [15] and the state-of-the-art RREA model [22] in this work for empirical evaluation. To validate the effectiveness of LIME, we create a large EA dataset FB_DBP_2M with millions of entities and tens of millions of facts. Experimental results demonstrate that LIME can effectively handle EA at scale while remaining reasonably effective and efficient. We also compare LIME against state-of-the-art solutions on three mainstream datasets, showing that LIME can achieve promising results even on small-scale datasets.

Contributions The main contributions of this chapter are the following:

- We identify the scalability issue in state-of-the-art EA approaches and propose an EA framework LIME to deal with large-scale entity alignment.
- We propose seed-oriented bidirectional graph partition strategies to partition large-scale KG pairs into smaller ones, where the alignment process is then conducted.
- We propose a reciprocal alignment inference strategy that models and integrates the bidirectional preferences of entities when inferring alignment results.
- We introduce two variants of reciprocal alignment inference that increase its scalability while incurring a small decrease in performance.
- Our proposed model, LIME, is generic and can be applied to existing EA models to enhance their ability to handle large-scale entity alignment.
- We demonstrate the effectiveness of our proposed model through a comprehensive experimental evaluation on popular entity alignment benchmarks and a newly constructed dataset with tens of millions of facts.

Organization In Sect. 5.2, we present the outline of LIME. In Sect. 5.3, we introduce the partition strategies. In Sect. 5.4, we introduce the reciprocal alignment inference strategy. In Sect. 5.5, we introduce the variants of reciprocal alignment inference. In Sects. 5.6 and 5.7, we introduce the experimental settings and results, respectively. In Sect. 5.8, we introduce related work, followed by conclusion in Sect. 5.9.

5.2 Framework

We present the overall framework of our proposal, LIME, in Fig. 5.2.

- To handle large-scale KGs, we begin by performing seed-oriented bidirectional partition (SBP) to partition the source and target KGs into multiple subgraph pairs with the aid of seed entity pairs.
- Subsequently, for each subgraph pair, we employ a KG structural learning model[2] to generate unified entity embeddings, enabling direct comparisons between entities from different KGs.
- Afterward, using the unified entity representations, we apply a reciprocal alignment inference strategy to model entity preferences on both sides and aggregate bidirectional preference information to generate alignment results. We also recognize that although reciprocal modeling achieves superior performance, it is computationally expensive in terms of time and memory. Therefore, we propose two alternative strategies to reduce the memory and time consumption at a slight cost of effectiveness: no-ranking aggregation and progressive blocking.
- Moreover, we introduce an iterative version of LIME to enhance the partition performance by incorporating confident alignment results from the previous round with the seed entity pairs. This iterative process leads to gradual improvements in both partitioning and alignment results.

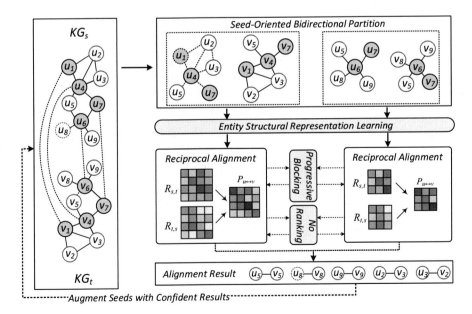

Fig. 5.2 The framework of our proposal. The entities in gray represent the seed entities. The corresponding seed entities are connected by dotted lines in the left of the figure

---

[2] Notice that LIME is agnostic to the choice of structural learning models.

5.3   Partition Strategies for Entity Alignment

To handle large-scale input KGs, a common approach is to partition the KGs and parallelize the computation across a distributed cluster of machines [3]. In this work, we adopt this approach and propose to partition KGs into smaller subgraphs, align entities in each subgraph pair, and aggregate the alignment results in each partition to produce the final aligned entity pairs.

We leverage the commonly used graph partition tool, METIS [13], as the basic partition strategy. The algorithms in METIS are based on multilevel graph partitioning [12, 14], which reduces the graph size by collapsing vertices and edges, partitioning the smaller graph, and then uncoarsening it to construct a partition for the original graph. The aim is to create a balanced vertex partition that equitably divides the set of vertices into multiple partitions while minimizing the number of edges spanning the partitions. However, in the case of EA, there are two separate graphs at scale and a small number of seed entity pairs connecting them. The two graphs are interlinked by the seed entity pairs and can be considered as one and forwarded to METIS for partitioning. Indeed, this approach is likely to generate subgraphs that only contain source or target entities, which is contrary to the goal of EA that aims to identify equivalent entities between KGs. Therefore, we use seed-oriented graph partition strategies in this work.

In this section, we first introduce the seed-oriented *unidirectional* graph partition strategy as the baseline model. Then, we describe our proposed bidirectional partition framework and its iterative variants.

5.3.1   Seed-Oriented Unidirectional Graph Partition

Unidirectional graph partition strategies for EA conduct only one-way partition (e.g., source-to-target) of KG pairs using the seed entity pairs. Formally, they partition the source KG $\mathcal{KG}_s$ and target KG $\mathcal{KG}_t$ into k subgraph pairs $\Phi = \{C_1, C_2, \ldots, C_k\}$, where each subgraph pair $C_i = \{\mathcal{KG}_s^i, \mathcal{KG}_t^i, S^i\}$ contains a pair of source subgraph $\mathcal{KG}_s^i$ and target subgraph $\mathcal{KG}_t^i$, as well as a number of seed entity pairs S^i connecting the subgraphs. Specifically, in this work, we adopt a state-of-the-art unidirectional partition strategy CPS [10] as the baseline model.

CPS first directly partitions the source KG into k subgraphs $\Phi_s = \{\mathcal{KG}_s^1, \ldots, \mathcal{KG}_s^k\}$ using METIS. Each source subgraph $\mathcal{KG}_s^i$ contains ε_i source entities $S_s^i = \{u_1^i, \ldots, u_{\varepsilon_i}^i\}$ from the seed entity pairs S. To partition the target KG ($\mathcal{KG}_t$), we still use METIS, but with some modifications: (1) we assign higher weights to edges among seed target entities whose corresponding source entities are in the same subgraph. This encourages METIS to place these seed target entities in the same subgraph while retaining the overall KG structure; and (2) we assign edges among seed target entities whose corresponding source entities are from different subgraphs, say S_s^i and S_s^j, with weight 0. This discourages

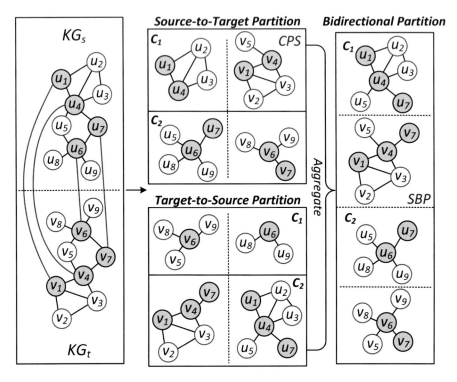

Fig. 5.3 Illustration of the partition process. In each box, the solid line separates different subgraph pairs, while the dotted line differentiates the source subgraphs from the target ones

placing these seed target entities in the same subgraph as their corresponding seed source entities are not in the same subgraph. Partitioning the target KG also results in k subgraphs $\Phi_t = \{\mathcal{KG}_t^1, \ldots, \mathcal{KG}_t^k\}$. Then, for each source subgraph $\mathcal{KG}_s^i$, it retrieves the target subgraph $\mathcal{KG}_t^*$ that possesses the largest number of target entities S_t^* corresponding to seed source entities S_s^i in $\mathcal{KG}_s^i$ and considers them as a subgraph pair $C_i = \{\mathcal{KG}_s^i, \mathcal{KG}_t^*, S^*\}$, where S^* refers to the links connecting S_t^* and S_s^i. We illustrate this process using the following example.

Example As shown in Fig. 5.3, there are two KGs to be aligned (i.e., $\mathcal{KG}_s$ and $\mathcal{KG}_t$), where the colored lines denote the links in seed entity pairs, and the seed entities are also represented in gray. The entities with the same subscripts are equivalent.

The proposed CPS conducts a one-off source-to-target partition. It first partitions $\mathcal{KG}_s$, resulting in two source subgraphs shown in the left part of the box. These subgraphs consist of $\{u_1, u_2, u_3, u_4\}$ and $\{u_5, u_6, u_7, u_8, u_9\}$,

(continued)

respectively. Next, when partitioning $\mathcal{KG}_t$, it increases the weight of the edge between v_1 and v_4 (resp., v_6 and v_7) since the seed source entities u_1 and u_4 (resp., u_6 and u_7) are in the same subgraph. Additionally, it sets the weight of the edge between v_4 and v_7 to 0. $\mathcal{KG}_t$ is thus partitioned into two subgraphs shown in the right part of the box, which consist of $\{v_1, v_2, v_3, v_4, v_5\}$ and $\{v_6, v_7, v_8, v_9\}$, respectively. Finally, using the seed entity pairs as anchors, it generates two subgraph pairs, i.e., C_1 and C_2.

5.3.2 Bidirectional Graph Partition

It is observed that in unidirectional partition strategies like CPS, the partition of the source KG can preserve its original structure well. However, the partition of the target KG is limited by the goal of retaining the seed entity pairs, which may lead to the destruction of the KG structure to some extent. As a solution, we propose a seed-oriented bidirectional graph partition framework, called SBP. The SBP framework first conducts the source-to-target partition using any unidirectional strategy, resulting in a set of source subgraphs (Φ_s^0) and a set of target subgraphs (Φ_t^0). Then, it conducts the partition process reversely, obtaining another set of source subgraphs (Φ_s^1) and target subgraphs (Φ_t^1). Next, it identifies and combines corresponding source subgraphs in Φ_s^0 and Φ_s^1, resulting in the aggregated set of source subgraphs (Φ_s). Similarly, it generates the aggregated set of target subgraphs (Φ_t). Finally, for each source subgraph ($\mathcal{KG}_s^i \in \Phi_s$), it retrieves the target subgraph ($\mathcal{KG}_t^* \in \Phi_t$) that possesses the largest number of seed target entities (S^*) corresponding to seed source entities in $\mathcal{KG}_s^i$. It considers them as a subgraph pair ($C_i = \{\mathcal{KG}_s^i, \mathcal{KG}_t^*, S^*\}$) for alignment. The detailed process is presented in Algorithm 1 and the following example.

Example Continuing with the previous example, the SBP framework conducts the target-to-source partition, resulting in two target subgraphs comprising $\{v_1, v_2, v_3, v_4, v_7\}$ and $\{v_5, v_6, v_8, v_9\}$ and two source subgraphs comprising $\{u_1, u_2, u_3, u_4, u_5, u_7\}$ and $\{u_6, u_8, u_9\}$. Next, it identifies and combines corresponding source and target subgraphs generated by the source-to-target and target-to-source partition. For instance, based on the number of overlapping source seed entities, it identifies that the source subgraph comprising $\{u_1, u_2, u_3, u_4\}$ (resp., $\{u_5, u_6, u_7, u_8, u_9\}$) generated by the source-to-target partition and the source subgraph comprising $\{u_1, u_2, u_3, u_4, u_5, u_7\}$ (resp., $\{u_6, u_8, u_9\}$) generated by the target-to-source partition are corresponding. It

(continued)

Algorithm 1: Bidirectional graph partition (SBP)

Input : $\mathcal{KG}_s$: source KG; $\mathcal{KG}_t$: target KG; S:seed pairs.
Output : $\Phi = \{C_1, \ldots, C_k\}$: k subgraph pairs.
1 Conduct source-to-target partition using any unidirectional partition strategy (e.g., CPS).
  Obtain Φ_s^0 and Φ_t^0;
2 Conduct target-to-source partition using any unidirectional partition strategy (e.g., CPS).
  Obtain Φ_s^1 and Φ_t^1;
3 $\Phi_s \leftarrow \emptyset$;
4 **foreach** $\mathcal{KG}_s^i \in \Phi_s^0$ **do**
5   Identify the source subgraph $\mathcal{KG}_s^* \in \Phi_s^1$ that has the largest number of overlapping seed entities with $\mathcal{KG}_s^i$;
6     $\Phi_s \leftarrow \Phi_s \cup \{\mathcal{KG}_s^i \cup \mathcal{KG}_s^*\}$;
7 $\Phi_t \leftarrow \emptyset$;
8 **foreach** $\mathcal{KG}_t^i \in \Phi_t^0$ **do**
9   Identify the target subgraph $\mathcal{KG}_t^* \in \Phi_t^1$ that has the largest number of overlapping seed entities with $\mathcal{KG}_t^i$;
10     $\Phi_t \leftarrow \Phi_t \cup \{\mathcal{KG}_t^i \cup \mathcal{KG}_t^*\}$;
11 $\Phi \leftarrow \emptyset$;
12 **foreach** $\mathcal{KG}_s^i \in \Phi_s$ **do**
13   Retrieve the target subgraph $\mathcal{KG}_t^* \in \Phi_t$ that possesses the largest number of target seed entities S_t^* corresponding to the seed source entities $S_s^i \subset \mathcal{KG}_s^i$;
14     $C_i \leftarrow \{\mathcal{KG}_s^i, \mathcal{KG}_t^*, S^*\}$;
15     $\Phi \leftarrow \Phi \cup \{C_i\}$;
16 **return** Φ.

---

combines them to generate the aggregated subgraph $\{u_1, u_2, u_3, u_4, u_5, u_7\}$ (resp., $\{u_5, u_6, u_7, u_8, u_9\}$). The target subgraphs are aggregated in the same way. Finally, using the seed entity pairs as anchors, it generates two subgraph pairs, as shown in the rightmost box.

As shown in Fig. 5.3, in the partition results of CPS, equivalent entities may be placed in different subgraph pairs, such as u_5 and v_5. The SBP framework can effectively mitigate this issue by conducting bidirectional partitions and aggregating the results. Hence, while the partition results of the SBP framework may include redundant entities that exist in multiple subgraph pairs, it can still effectively decrease the instances where equivalent entities are allocated to different subgraph pairs.

Merits of Bidirectional Partitioning Noteworthily, Φ_s^0 (resp., Φ_t^1) is generated with the aim of preserving the original KG structure, while Φ_s^1 (resp., Φ_t^0) is generated with the aim of both retaining the links and preserving the original KG structure. Consequently, the integration of subgraphs in Φ_s^0 (and Φ_t^1) with Φ_s^1 (and

Φ_t^0) results in aggregated subgraphs in Φ_s and Φ_t that have a more comprehensive structure and a greater number of seed entities pointing to the subgraphs in the opposite side. This can ultimately lead to more precise alignment outcomes.

Moreover, unlike Φ_s^0, Φ_s^1, Φ_t^0, or Φ_t^1, where the subgraphs do not have common entities, the subgraphs in Φ_s and Φ_t overlap. This is comparable to the concept of redundancy-based methods in traditional entity resolution (ER) blocking techniques, where an entity can be assigned to multiple blocks [26]. This is because the partitioning process may unavoidably assign equivalent entities to different subgraph pairs, which limits the upper bound of the alignment performance (as the alignment is only performed within each subgraph pair). However, this upper bound can be raised through bidirectional partitioning, which assigns an entity to multiple subgraph pairs. This is empirically validated in Sect. 5.7.3.

Integration of Subgraph-Wise Alignment Results As previously mentioned, the partition results produced by unidirectional strategies do not have redundancies, and therefore, the alignment outcomes can be obtained by directly merging subgraph-wise alignment results. However, since the subgraph pairs generated by SBP may contain overlapping entities, an additional result aggregation module is necessary to resolve any potential conflicts in the alignment outcomes. To address this, we adopt a straightforward *voting* strategy. Specifically, for the source entity aligned to multiple target entities generated by different subgraph pairs, we choose the target entity with the highest number of "votes" from the subgraph pairs as the final alignment outcome. If multiple target entities have the same highest vote, we select the one with the lowest mutual preference rank (explained in Sect. 5.4.3) as the match.

5.3.3 Iterative Bidirectional Graph Partition

It is clear that the one-off partitioning approach tends to generate inaccurate partition results, where equivalent entities may be placed in different subgraph pairs, and the original KG structure information could also be partially lost. To address this issue, we propose an iterative framework called I-SBP, which performs the partitioning process for γ rounds based on the signals provided by the previous round. Specifically, in each iteration, we partition the KG into k subgraph pairs using SBP and perform entity alignment within each subgraph pair (detailed in the next section). We then aggregate the subgraph-wise alignment results to generate the final aligned entity pairs. Since the final alignment results include confident entity pairs, which can be considered as pseudo seeds according to previous studies [33, 45], we select these entity pairs using the bidirectional nearest neighbor search in [45] and add them to the seed entity pairs S to aid the partition in the next round. The process is detailed in Algorithm 2.

Algorithm 2: Iterative SBP (I-SBP)

Input : $\mathcal{KG}_s$: source KG; $\mathcal{KG}_t$: target KG; S:seed pairs.
Output : $\Phi = \{C_1, C_2, \ldots, C_k\}$: k subgraph pairs.
1 $r \leftarrow 0$;
2 **while** $r < \gamma$ **do**
3   Obtain the $\Phi = \{C_1, C_2, \ldots, C_k\}$ via Algorithm 1;
4   Perform alignment in each subgraph pair and generate results;
5   Select confident alignment results and add them to S;
6     $r \leftarrow r + 1$;
7 **return** Φ.

5.3.4 Complexity Analysis

The time complexity of SBP is roughly double that of the unidirectional partition strategy it employs, while the time complexity of I-SBP is approximately γ times that of SBP. We use CPS as the unidirectional partition strategy in this study, and its time complexity is $O(|S| + \frac{(2k-1)|S|^2}{k^2} + |E_s| + |E_t| + |T_s| + |T_t| + k \log(k))$ [10], where $|E_s|$ (and $|E_t|$) and $|T_s|$ (and $|T_t|$) represent the number of entities and triples in the source (and target) KG, respectively, $|S|$ refers to the number of seed entity pairs, and k denotes the number of subgraph pairs.

Regarding space complexity, most unidirectional partition strategies need to store two knowledge graphs (KGs) simultaneously. However, for the SBP algorithm, bidirectional partitions are required, which necessitates the storage of four KGs. The space complexity of I-SBP is similar to that of SBP. In general, the space complexity of the partition process is determined by the size of the knowledge graphs involved.

5.3.5 Discussion

It is important to note that partition strategies are used to divide large-scale knowledge graph pairs into smaller ones so that state-of-the-art deep learning-based methods can be used to identify equivalent entities. However, the partition process can reduce the alignment performance as equivalent entities can be placed into different subgraph pairs. While this issue can be mitigated by improving partition strategies, it cannot be entirely avoided. Therefore, when dealing with small- or medium-sized datasets such as current entity alignment benchmarks, it may not be worthwhile to use partition strategies since partitioning would not significantly reduce computational costs while compromising alignment accuracy. This is also supported by empirical evidence from experiments conducted on the DWY100K dataset. Whether to use partition strategies ultimately depends on the alignment goal, i.e., efficiency or effectiveness. In this work, we follow previous works and do not employ partition strategies when dealing with small- or medium-sized EA datasets, except for the analysis of partition strategies.

5.4 Reciprocal Alignment Inference

After partitioning large-scale knowledge graph pairs into smaller ones, we perform alignment on each subgraph pair and combine the alignment results. In this subsection, we provide a brief overview of the representation learning process. Additionally, we propose a reciprocal inference strategy (illustrated in Fig. 5.4) that takes into account the mutual interactions between bidirectional alignments to enhance the alignment inference process. This strategy allows us to capture reciprocal interactions and improve alignment inference.

5.4.1 Entity Structural Representation Learning

The entity structural representation learning phase aims to model the structural characteristics of entities and project them from different knowledge graphs into a unified embedding space. In this space, the similarity between entities can be directly inferred by comparing their structural embeddings. Most state-of-the-art EA solutions focus on improving this phase by designing advanced structural representation learning models. However, our focus in this work is to enhance the alignment inference process and the capability of EA models to handle large-scale datasets. As such, our proposed model, LIME, is agnostic to the choice of structural learning models. We adopt a state-of-the-art embedding learning model

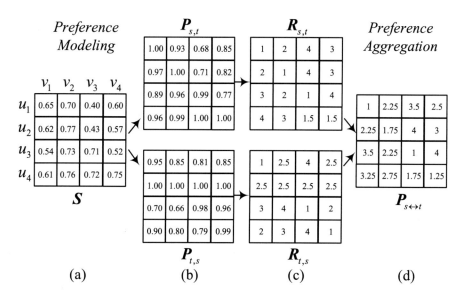

Fig. 5.4 An example of the preference modeling and aggregation. (**a**) similarity matrix, (**b**) preference matrix, (**c**) ranking matrix (**d**) reciprocal matrix

for EA, RREA [22], which reflects entity representations along different relational hyperplanes to construct relation-specific entity embeddings for alignment. More model and implementation details can be found in the original paper. Besides, to demonstrate that LIME is generic and can be applied to existing representation learning models, we also adopt the most commonly used model in the EA literature, GCN [15, 38], as the baseline model. Relevant experimental evaluations can be found in Sect. 5.7.

5.4.2 Preference Modeling

Once we have obtained the unified entity representations, we can infer the alignment results based on entity preferences. Specifically, for each source entity, we predict its most preferred target entity as its equivalent entity.

Direct Alignment Inference Previous studies only considered the similarity between entity representations to model entity preferences. We refer to this as *direct alignment inference*. Given an entity pair $(u, v), u \in \mathcal{E}_s, v \in \mathcal{E}_t$, their similarity score is denoted as $sim(\boldsymbol{u}, \boldsymbol{v})$,[3] where $\boldsymbol{u}$ and $\boldsymbol{v}$ are the entity embeddings of u and v, respectively. The corresponding similarity matrix is denoted as S. For direct alignment inference, the preference score of u toward v is defined as:

$$p_{u,v} = sim(\boldsymbol{u}, \boldsymbol{v}). \tag{5.1}$$

According to this definition, the preference score of u toward v is the same as the preference score of v toward u. Therefore, we have $p_{u,v} = p_{v,u}$, since similarity measures are usually symmetric and do not differentiate between the two input elements.

Reciprocal Preference Modeling We believe that to accurately model entity preferences, an entity's preference score toward another entity should also consider the likelihood of a match between them. For instance, as can be observed from Fig. 5.1, for [B. Dessner]$_{en}$, despite the high similarity score, it might have a low preference toward [A. Dessner]$_{es}$, since in its view, they are less likely to form a match (considering that [A. Dessner]$_{es}$ has a higher similarity with [A. Dessner]$_{en}$). Theoretically, a source (target) entity would prefer the target (source) entities that have high similarities with it *and* meanwhile low similarities with other source (target) entities. In this connection, we define the preference score of u toward v as:

$$p_{u,v} = sim(\boldsymbol{u}, \boldsymbol{v}) - \max\{sim(\boldsymbol{v}, \boldsymbol{u}'), u' \in \mathcal{E}_s\} + 1, \tag{5.2}$$

---

[3] The similarity measure *sim* is usually chosen from cosine similarity [32–34], Euclidean distance [6, 44, 51], or Manhattan distance [38–40].

where $0 \leq p_{u,v} \leq 1$, and a larger $p_{u,v}$ denotes a higher degree of preference. The preference score of v toward u is defined similarly.

Our definition of the preference score for an entity toward another entity is composed of three elements. The first element represents the similarity score between the two entities, while the second element represents the highest similarity score that the target entity has with all the source entities. Intuitively, u would prefer v more if their similarity score $sim(u, v)$ is close (ideally, equal) to the highest similarity score that v has, i.e., $\max\{sim(v, u'), u' \in \mathcal{E}_s\}$. Hence, we subtract the first element from the second element. If the difference is close to 0, it shows that u is satisfied with v. To make the preference value positive, we add the third element, i.e., 1.

Our definition of the preference score takes into account the alignment in the reverse direction (i.e., the preference of the target entity toward the source entity), which is naturally incorporated into an entity's preference modeling. Moreover, $p_{u,v}$ is not necessarily equal to $p_{v,u}$, since the preference score encodes the alignment information at the entity level (rather than the pairwise level as in Eq. (5.1)). We denote the matrix forms of the source-to-target and target-to-source preference scores as $P_{s,t}$ and $P_{t,s}$, respectively, and in general $P_{s,t} \neq P_{t,s}^{\top}$, where $P_{t,s}^{\top}$ is the transpose of $P_{t,s}$.

5.4.3   Preference Aggregation

The preference scores only reflect the preferences in one direction, and an optimal alignment result should consider the preference scores in both directions. Hence, we propose to aggregate the unidirectional preferences. More specifically, we first convert the preference matrix P into the ranking matrix R. The elements in each row of $P_{s,t}$ and $P_{t,s}$ are ranked descendingly according to their values, resulting in $R_{s,t}$ and $R_{t,s}$, respectively.[4] Each element in the ranking matrix R represents the rank of the corresponding preference score, where a lower rank indicates a higher preference value. As thus, the ranking matrices can also encode the preference information.

The primary objective of transforming scores into ranks is to magnify the disparities between the scores. As we combine the source-to-target and target-to-source matrices to capture shared preferences, the small differences in scores on one side may be easily overlooked after the aggregation with information on the other side. Transforming scores into ranks allows us to preserve and integrate such differences into the ultimate mutual preference.

---

[4] For tied elements, we follow the common practice and denote their ranks as the average of the ranks that would have been assigned to them.

Afterward, we combine the two ranking matrices to capture the mutual preferences of entities and create the corresponding preference matrix:

$$P_{s \leftrightarrow t} = \phi\left(R_{s,t}, R_{t,s}^{\top}\right), \qquad (5.3)$$

where ϕ is an aggregation function, which can be chosen from any mean operators, cross-ratio uniform [43], or other viable methods [25]. For this study, we use the arithmetic mean, which remains impartial and presents both entities' preferences toward each other precisely, without showing any inclination toward a higher or lower rank [23, 25]. The reciprocal matrix contains elements denoted by $p_{u \leftrightarrow v}$, which indicates the degree of mutual preference between a pair of entities (u and v). The lower the value of $p_{u \leftrightarrow v}$, the higher the level of preference between the two entities.

Algorithm 3: reciprocal_inference($\mathcal{E}_s, \mathcal{E}_t, S$)

Input   : $\mathcal{E}_s$ and $\mathcal{E}_t$: the entity sets in the two KGs; S: the similarity matrix;
Output : $\mathcal{M}$: the set of aligned entity pairs.
1 **for** $u \in \mathcal{E}_s$ **do**
2 | **for** $v \in \mathcal{E}_t$ **do**
3 | | Calculate $p_{u,v}$ and $p_{v,u}$ (cf. Eq. (5.2));

4 Aggregate and collect the preference scores in $P_{s,t}$ and $P_{t,s}$;
5 **foreach** row in $P_{s,t}$ and $P_{t,s}$ **do**
6 | Rank the elements in each row descendingly;
7 | **if** the values of the elements are the same **then**
8 | | Denote their ranks as the average of the ranks that would have been assigned to them;

9 Aggregate the ranking matrices $R_{s,t}$ and $R_{t,s}$ to produce reciprocal preference matrix $P_{s \leftrightarrow t}$ (cf. Eq. (5.3));
10 **for** $u \in \mathcal{E}_s$ **do**
11 | **if** $p_{u \leftrightarrow v^*} \leq p_{u \leftrightarrow v'}, \forall v' \in \mathcal{E}_t$ **then**
12 | | $\mathcal{M} \leftarrow \mathcal{M} \cup \{(u, v^*)\}$;

13 **return** $\mathcal{M}$;

Algorithm 3 provides the details of the reciprocal alignment inference process. We also use Example 1 to further illustrate the process.

Example 1 As shown in Fig. 5.4, there are a total of four source entities (u_1, u_2, u_3, u_4) and four target entities (v_1, v_2, v_3, v_4). In S, $P_{s,t}$, $R_{s,t}$, and $P_{s \leftrightarrow t}$, the rows correspond to the source entities and the columns correspond to the target entities, while in $P_{t,s}$ and $R_{t,s}$, the rows correspond to the target entities and the columns correspond to the source entities. The entities with the same subscripts are equivalent.

(a): The similarity scores in matrix S are computed by using cosine similarity with entity embeddings. If we consider these similarity scores as the entity preferences and align each source entity to its most preferred target entity, the results, such as (u_1, v_2), (u_2, v_2), (u_3, v_2), and (u_4, v_2), will only contain one correct match in this example.

(b): Using Eq. (5.2), we calculate the preference scores and obtain the preference matrices $P_{s,t}$ and $P_{t,s}$. Based on the preference scores, if we align each source entity to its most preferred target entity, the results (u_1, v_1), (u_2, v_2), and (u_3, v_3) are correct. However, for u_4, both v_3 and v_4 are likely to be the correct match. Additionally, if we align each target entity to its most preferred source entity, it is difficult to determine the result for v_2.

(c): We convert the preference matrices $P_{s,t}$ and $P_{t,s}$ into the ranking matrices $R_{s,t}$ and $R_{t,s}$, respectively.

(d): We can aggregate the two ranking matrices using the arithmetic mean. Based on the reciprocal preference matrix $P_{s \leftrightarrow t}$, if we align each source (target) entity to its most preferred source (target) entity, all the results will be correct.

$\square$

Discussion Some might reckon that Eq. (5.2) is similar to the definition of cross-domain similarity local scaling (CSLS) [16], a metric that is proposed to mitigate the hubness issue during the nearest neighbor search. However, CSLS subtracts the average of the top-n highest similarity scores of *both source and target* entities from the pairwise similarity, resulting in a score that is still at the pairwise level and cannot fully characterize the preference of each entity. On the other hand, our proposed entity-level preference measure can better reflect the individual preferences of entities, and the integrated reciprocal preference matrix leads to more accurate alignment results, as demonstrated in Sect. 5.7.4.

5.4.4 Correctness Analysis

The optimal solution for alignment inference is to correctly identify all entity pairs. For example, in Fig. 5.4, where there are four source entities (u_1, u_2, u_3, u_4), four target entities (v_1, v_2, v_3, v_4), and the similarity matrix S, the optimal solution is $\mathcal{M} = \{(u_1, v_1), (u_2, v_2), (u_3, v_3), (u_4, v_4)\}$.

Nevertheless, the ability of Algorithm 3 to attain a correct or optimal solution depends on the input similarity matrix S, which is generated by the deep learning-based representation learning process that captures the relatedness among entities. In the worst-case scenario, where the representation learning process fails to learn anything useful and the similarity matrix is composed of 0s, Algorithm 3 or any alignment inference strategy, such as direct alignment inference, would produce results that are full of wrongly aligned entity pairs. However, if the similarity matrix is accurate (i.e., for each ground-truth entity pair (u, v), u has a higher similarity score with v than the rest of the source entities, and likewise, v has a higher similarity score with u than the rest of the target entities), Algorithm 3 can find the correct solution, as proven in Proof.

Proof We prove that, given an accurate similarity matrix where for each ground-truth entity pair (u, v):

$$u = \arg\max\{sim(\boldsymbol{v}, \boldsymbol{u}'), u' \in \mathcal{E}_s\};$$
$$v = \arg\max\{sim(\boldsymbol{u}, \boldsymbol{v}'), v' \in \mathcal{E}_t\},$$

the reciprocal inference algorithm could accurately identify these entity pairs.

Without loss of generality, we consider the ground-truth entity pair (u, v). The following proof also applies to the rest of the ground-truth entity pairs.

First, we can derive that $p_{u,v} = 1$ and $p_{v,u} = 1$ according to Eq. (5.2). Further, we can derive that:

$$p_{u,v} = \max\{p_{u,v'}, v' \in \mathcal{E}_t\}; \quad p_{v,u} = \max\{p_{v,u'}, u' \in \mathcal{E}_s\}.$$

After converting the scores into ranks, we have:

$$r_{u,v} = \min\{r_{u,v'}, v' \in \mathcal{E}_t\}; \quad r_{v,u} = \min\{r_{v,u'}, u' \in \mathcal{E}_s\},$$

where r denotes the rank value. Next, after aggregating with arithmetic mean using Eq. (5.3), we can derive that:

$$p_{u\leftrightarrow v} = \min\{p_{u\leftrightarrow v'}, v' \in \mathcal{E}_t\}; \quad p_{v\leftrightarrow u} = \min\{p_{v\leftrightarrow u'}, u' \in \mathcal{E}_s\}.$$

Finally, according to Line 10 to Line 12 in Algorithm 3, u and v would be aligned by reciprocal alignment inference.   □

Therefore, the main challenge lies in obtaining an accurate similarity matrix. However, in most cases, the similarity matrix is likely to be inaccurate, as the representation learning process cannot guarantee to learn high-quality entity representations for generating an accurate similarity matrix. Thus, we categorize the similarity scores of the ground-truth entity pairs into four cases and discuss the performance of our proposed reciprocal alignment inference and the direct inference (baseline model) under these circumstances in the appendix. Empirically, reciprocal alignment inference achieves much better results than direct inference.

5.4.5   Complexity Analysis

Regarding the worst time complexity of Algorithm 3, the preference modeling process (Lines 1–3) requires $O(n^2) + O(2n^2) + O(2n^2)$ as we can calculate the highest similarity scores outside of the loops, the ranking process (Lines 5–8) requires $O(2n \times n \lg n)$, the aggregation process (Line 9) requires $O(n^2)$, and the matching process (Lines 10–12) requires $O(n^2)$, where n denotes the number of entities in a KG. Overall, the time complexity of Algorithm 3 is $O(n^2 \lg n)$. Notably, the time complexity of the direct alignment inference strategy is $O(n^2)$. Our proposed reciprocal alignment inference has a higher time complexity than direct inference as it includes an additional ranking process that converts the preference scores into ranks. However, the process of ranking is crucial for enhancing the alignment performance and will be confirmed through experimentation.

The primary factor contributing to the space complexity of LIME is the reciprocal alignment inference stage, specifically the computation of the similarity, preference, and ranking matrices. In contrast, the direct alignment inference approach only requires computing the similarity matrix with a size of $n \times n$, where n represents the number of entities. In our reciprocal modeling strategy, we remove the matrices once they are no longer necessary to decrease memory usage, and only up to three matrices are present at any given time. Thus, our model's maximum memory consumption is three times that of the direct alignment inference.

5.5 Variants of Reciprocal Alignment Inference

As discussed in Sect. 5.4.5, incorporating the reciprocal preferences of entities into the model requires a greater amount of memory and time compared to the direct alignment strategy, as a result of computing the preference and ranking matrices. Consequently, in this section, we propose two alternative methods to minimize the memory and time usage associated with the reciprocal modeling.

5.5.1 No-Ranking Aggregation

The complexity analysis in Sect. 5.4.5 has identified that the increased time complexity is primarily due to the calculation of preference score rankings. Therefore, we propose a no-ranking aggregation strategy in order to approximate reciprocal alignment inference, which eliminates the ranking process and instead directly aggregates $P_{s,t}$ and $P_{t,s}$ to produce the reciprocal preference matrix $P_{s \leftrightarrow t}$.

5.5.2 Progressive Blocking

To further reduce the time and space requirements of reciprocal alignment inference, we propose a method to decrease the value of n. We introduce a progressive blocking method that partitions the entities into smaller blocks and infers the alignment results at the block level. The algorithm for this method is presented in Algorithm 4 and the process is illustrated in Fig. 5.5.

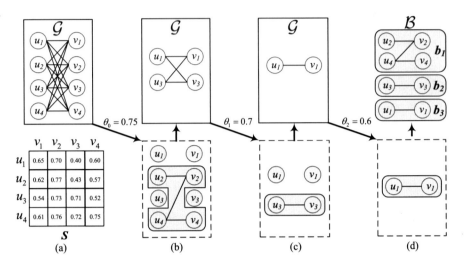

Fig. 5.5 An example of the progressive blocking process. The shape in gray denotes a block. For instance, there are five blocks in (**b**), i.e., $\{u_2, v_2, u_4, v_4\}$, $\{u_1\}$, $\{v_1\}$, and $\{u_3\}$, and $\{v_3\}$. (**a**) The unified graph and the reciprocal matrix, (**b**) first round of blocking, (**c**) second round of blocking, (**d**) third found of blocking

Difference from the Graph Partition Strategies It is important to note that the progressive blocking process and the graph partition strategies presented in Sect. 5.3 are distinct, despite both being methods for dividing large graphs into smaller ones. The input to the graph partition strategies is a KG, and the goal is to partition it into smaller subgraphs while preserving the original KG structure. In contrast, the input to the progressive blocking method is a bipartite graph with nodes representing source and target entities to be aligned and edges representing pairwise connections between them. The aim is to divide the bipartite graph into smaller blocks, where alignment can be inferred within a smaller search space. Consequently, when aligning large KG pairs, we first conduct graph partitioning to divide the KGs into smaller subgraphs. For each small KG pair, we learn entity structural representations and reciprocally infer the alignment results, where the progressive blocking method can be used to reduce the time and memory costs of reciprocal inference.

One-Off Blocking To provide more detail, the inputs to the progressive blocking process include the unified graph $\mathcal{G}$, which contains entities from both source and target KGs and their pairwise connections; the similarity matrix S, which encodes the pairwise similarities between source and target entities; and Θ, which is the set of given thresholds (hyper-parameters). First, the blocking process begins by removing the connections between a source entity u and a target entity v if the similarity score between them $sim(u, v)$ is lower than a predefined threshold $\theta \in \Theta$. This division creates different blocks of source and target entities, as illustrated in Fig. 5.5b. After obtaining the blocks, we perform reciprocal entity alignment on the entities within each block and aggregate the results from different blocks to obtain the overall alignment performance.

It is important to note that setting a small value for θ will result in most connections remaining, and most entities remaining in the same block. Therefore, the threshold is typically set to a relatively large value to ensure that the entities are effectively divided into appropriate blocks. However, this blocking process may still produce isolated blocks containing only a single entity, as depicted in Fig. 5.5b. We have found that these isolated blocks can represent a significant portion of the overall entities. One intuitive approach to handling these isolated entities is to gather them together and place them in the same block. However, this block would likely be large in size, and reciprocally aligning the entities within it would still require a significant amount of memory (as empirically validated in Sect. 5.7.5).

Algorithm 4: progressive_blocking($\mathcal{G}$, S, Θ)

Input   : $\mathcal{G}$: the unified graph containing entities from the two KGs and their pair-wise connections; S: the similarity matrix; Θ: the set of given thresholds;

Output : $\mathcal{B}$: a set of blocks.

1 **foreach** $\theta \in \Theta$ **do**
2 $\quad$ Remove from $\mathcal{G}$ the connections with similarity scores lower than θ;
3 $\quad$ $C \leftarrow$ compute the connected components of $\mathcal{G}$;
4 $\quad$ $\mathcal{G} \leftarrow \emptyset$;
5 $\quad$ **foreach** $c \in C$ **do**
6 $\quad\quad$ **if** $|c| > 1$ **then**
7 $\quad\quad\quad$ $\mathcal{B} \leftarrow \mathcal{B} \cup \{c\}$;
8 $\quad\quad$ **else**
9 $\quad\quad\quad$ $\mathcal{G} \leftarrow \mathcal{G} \cup \{c\}$;
10 $\quad$ Restore the edges associated with $\mathcal{G}$;   /* previously removed in Line 2 */
11 **return** $\mathcal{B}$;

Progressive Blocking To address this issue, we propose a progressive blocking strategy. The strategy begins by removing connections between source and target entities in $\mathcal{G}$ with similarity scores lower than the threshold θ and computing the connected components of $\mathcal{G}$. Each connected component is considered as a block (Lines 2–3 in Algorithm 4). For each block in the block sets, if it contains more than one entity, it is added to the final set of blocks (Lines 6–7). We gather up the isolated entities (blocks) and place them into one block and restore the connections among the entities in this block, which forms the new unified graph $\mathcal{G}$ (Lines 9–10). Then, we choose the next θ (smaller than the previous one) from Θ and block $\mathcal{G}$ using the same strategy, i.e., removing connections with similarity scores lower than θ. As the threshold is lower than the previous one, some of the connections among these entities would remain, and these entities would be placed into different blocks. Similarly, there may still be isolated entities, and we can repeat the progressive blocking strategy to generate more non-isolated blocks by gathering up the isolated entities and adjusting the threshold. Finally, we obtain the final set of blocks (Line 11). We perform reciprocal entity alignment within each block and aggregate the individual results to attain the final alignment performance. A running example can be found in the following example.

The Benefits and Limitations of Progressive Blocking By applying the progressive blocking method, the memory and time costs of reciprocal alignment inference are significantly reduced, as the number of entities in each block is much smaller than in the original graph. This reduction is empirically validated in Table 5.2 in Sect. 5.7.1. However, the blocking process may partition equivalent entities into different blocks, which can negatively impact the alignment accuracy. This issue is further discussed and analyzed in Sect. 5.7.

> *Example* Continuing with the previous example, we now explain the progressive blocking process, which is also illustrated in Fig. 5.5.
>
> (a) The inputs include the unified graph G and the similarity matrix S.
> (b) We set the initial threshold θ_0 to 0.75 and remove the pairwise connections in G with similarity scores lower than θ_0. This results in five blocks, i.e., $\{u_2, v_2, u_4, v_4\}$, $\{u_1\}$, $\{v_1\}$, $\{u_3\}$, and $\{v_3\}$, among which the latter four blocks contain only one entity. We gather these entities and restore connections among them, resulting in G_0.
> (c) We lower the threshold and set it to $\theta_1 = 0.7$. We then remove the connections in G_0 with similarity scores lower than θ_1, which generates three blocks, $\{u_3, v_3\}$, $\{u_1\}$, and $\{v_1\}$. The latter two blocks are isolated, and we aggregate them to generate G_0.
> (d) Again, we lower the threshold to $\theta_2 = 0.6$. The similarity score between u_1 and v_1 is higher than this threshold and they form a block. Finally, we obtain three blocks, $\mathcal{B} = \{b_1, b_2, b_3\} = \{\{u_2, v_2, u_4, v_4\}, \{u_3, v_3\}, \{u_1, v_1\}\}$.

5.6 Experimental Settings

In this section, we introduce the experimental settings.

5.6.1 Dataset

Following previous works, we adopt three popular EA datasets for evaluation:

- DBP15K [32], which includes three cross-lingual KG pairs extracted from DBpedia [1], i.e., DBP15K$_{\text{ZH-EN}}$ (Chinese to English), DBP15K$_{\text{JA-EN}}$ (Japanese to English), and DBP15K$_{\text{FR-EN}}$ (French to English). Each KG pair comprises 15,000 aligned entity pairs and approximately 200,000 relational triples.
- SRPRS [11], which involves two cross-lingual KG pairs that are extracted from DBpedia, i.e., SRPRS$_{\text{EN-FR}}$ (English to French) and SRPRS$_{\text{EN-DE}}$ (English to German), and two mono-lingual datasets, i.e., SRPRS$_{\text{DBP-WD}}$ (DBpedia to Wikidata [37]) and SRPRS$_{\text{DBP-YG}}$ (DBpedia to YAGO [30]). Compared with DBP15K, the entity degree distribution in SRPRS is closer to the real-life distribution. Each KG pair comprises 15,000 aligned entity pairs and approximately 70,000 relational triples.
- DWY100K [33], which involves two mono-lingual KG pairs, i.e., DWY100K$_{\text{DBP-WD}}$ (DBpedia to Wikidata) and DWY100K$_{\text{DBP-YG}}$ (DBpedia to YAGO). Each KG pair comprises 100,000 aligned entity pairs and approximately 900,000 relational triples. Compared with DBP15K and SRPRS, the scale of DWY100K is much larger.

Table 5.1 presents a summary of the statistics of these datasets. We use 30% of the aligned pairs for training and 10% for validation.

Table 5.1 Statistics of the datasets used for evaluation

Dataset	KG pairs	#Triples	#Entities	#Relations	#Align.
DBP15K$_{\text{ZH-EN}}$	DBpedia(Chinese)	70,414	19,388	1,701	15,000
	DBpedia(English)	95,142	19,572	1,323	
DBP15K$_{\text{JA-EN}}$	DBpedia(Japanese)	77,214	19,814	1,299	15,000
	DBpedia(English)	93,484	19,780	1,153	
DBP15K$_{\text{FR-EN}}$	DBpedia(French)	105,998	19,661	903	15,000
	DBpedia(English)	115,722	19,993	1,208	
SRPRS$_{\text{EN-FR}}$	DBpedia(English)	36,508	15,000	221	15,000
	DBpedia(French)	33,532	15,000	177	
SRPRS$_{\text{EN-DE}}$	DBpedia(English)	38,363	15,000	222	15,000
	DBpedia(German)	37,377	15,000	120	
SRPRS$_{\text{DBP-WD}}$	DBpedia	38,421	15,000	253	15,000
	Wikidata	40,159	15,000	144	
SRPRS$_{\text{DBP-YG}}$	DBpedia	33,748	15,000	223	15,000
	YAGO3	36,569	15,000	30	
DWY100K$_{\text{DBP-WD}}$	DBpedia	463,294	100,000	330	100,000
	Wikidata	448,774	100,000	220	
DWY100K$_{\text{DBP-YG}}$	DBpedia	428,952	100,000	302	100,000
	YAGO3	502,563	100,000	31	
FB_DBP_2M	Freebase	13,502,306	2,349,253	4,901	2,349,253
	DBpedia	11,207,773	2,349,253	612	

5.6.2 Construction of a Large-Scale Dataset

To evaluate the scalability of EA, we created a new dataset with millions of entities by using DBpedia and Freebase as the source and target KGs, respectively. We obtain the gold standards, i.e., aligned entity pairs, from the external links between DBpedia and Freebase.[5] We extract the relational triples involving the entities in the external links from the respective KGs. We then extract the relational triples involving these entities from their respective KGs. To ensure the quality of the extracted triples, we follow the method proposed in a previous work [50]. We keep only the links whose source and target entities are involved in at least one triple in their respective KGs, and the entity sets are adjusted accordingly. As a result of this process, each KG contains over two million entities and tens of millions of triples. Table 5.1 presents the statistics of the newly constructed dataset.

5.6.3 Implementation Details

For the graph partition, we set the number of subgraph pairs k to 75 for FB_DBP_2M and 5 for DWY100K. For CPS, we adopt the same settings as the original paper. The number of rounds γ of I-SBP is set to 3. For the representation learning models RREA and GCN, we adopt the same settings as the original papers [22, 38]. We use cosine similarity to measure the similarity between entity embeddings. The reciprocal alignment inference stage does not require any additional parameters. Regarding the progressive blocking process, we conduct three rounds and set the thresholds Θ (hyper-parameters) to the 50th percentile (median), 25th percentile (the first quartile), and 1st percentile of the set of the largest similarity score of each source entity, respectively, which could be directly obtained given the similarity matrix. The main intuition behind this is that such settings can guarantee the thresholds are decreasing, and meanwhile the threshold values would not be too small as they are obtained from the set of the largest similarity scores of all source entities.

To compare with the approaches that leverage extra information, we incorporate entity names into our proposal. We directly adopt the strategies proposed in [46] to generate useful features from entity names for alignment. We do acknowledge that some methods [36, 44] use entity descriptions to improve the alignment performance significantly. However, we leave the integration of such information and comparison with these methods to future work, as it is outside the scope of this study. The source codes of LIME are publicly available at https://github.com/DexterZeng/LIME.

---

[5] https://www.dbpedia.org/blog/dbpedia-is-now-interlinked-with-freebase-links-to-opencyc-updated/.

5.6.4 Evaluation Metrics

As per convention [6], we adopt Hits@1 as the performance measure, which indicates the proportion of correct alignments. Unless otherwise specified, Hits@1 is represented in percentage. We omit the frequently used Hits@10 and mean reciprocal rank (MRR) metrics since (1) they are less important indicators as pointed out in previous works [6, 50] and (2) they show similar trends to Hits@1.

In addition, we assess the alignment methods based on their memory usage (in GB) and time consumption (in seconds).

5.6.5 Competing Methods

Our model is compared against 24 methods, which are categorized into two groups. The first category consists of methods that employ various embedding learning models to acquire valuable entity representations for alignment, such as the following:

- MTransE (2017) [6]: This work uses TransE to learn entity embeddings.
- RSNs (2019) [11]: This work integrates recurrent neural networks with residual learning to capture the long-term relational dependencies within and between KGs.
- MuGNN (2019) [4]: This work proposes a new multichannel graph neural network model that aims to learn alignment-focused embeddings for knowledge graphs by effectively encoding two KGs through multiple channels.
- KECG (2019) [17]: The aim of this research paper is to suggest a method for learning a knowledge embedding model and a cross-graph model together. The knowledge embedding model is responsible for encoding inner-graph relation-ships, while the cross-graph model improves entity embeddings by incorporating information from their neighbors.
- TransEdge (2019) [34]: It introduces a new embedding model that is focused on the edge or relation between entities. This model contextualizes the representation of relations by considering the specific pair of head and tail entities involved.
- MMEA (2019) [28]: This paper proposes to model the multi-mapping relations in KG for EA.
- AliNet (2020) [35]: It suggests an EA network that utilizes attention and gating mechanism to aggregate information from both direct and distant neighborhoods.
- MRAEA (2020) [21]: This work involves creating a model that generates cross-lingual entity embeddings by focusing on the node's incoming and outgoing neighbors, as well as the meta semantics of the relations it is connected to.
- SSP (2020) [24]: This work proposes to combine the global structure of the knowledge graph with relational triples specific to each entity for alignment.

- LDSD (2020) [5]: This paper proposes to capture both short-term variations and long-term interdependencies within knowledge graphs, with the goal of achieving better alignment.
- HyperKA (2020) [31]: This work puts forward a hyperbolic relational graph neural network to embed knowledge graphs, utilizing a hyperbolic transformation to capture associations between pieces of knowledge.
- RREA (2020) [22]: This work involves reflecting entity embeddings across various relational hyperplanes, in order to create relation-specific entity embeddings that can be utilized for alignment.

The second group of techniques makes use of data beyond the KG structure. This comprises the following:

- JAPE (2017) [32]: This work uses the attributes of entities to refine the structural information.
- GCN-Align (2018) [38]: This work utilizes GCN to produce entity embeddings, which are then merged with attribute embeddings to align entities present in separate KGs.
- RDGCN (2019) [39]: The proposed approach involves a dual-graph convolutional network that is capable of incorporating relation information through attentive interactions between a knowledge graph and its dual relation counterpart.
- HGCN (2019) [40]: This work proposes to learn entity and relation representations for EA jointly.
- GM-EHD-JEA (2020) [42]: This work presents two coordinated reasoning techniques that can effectively address the many-to-one problem encountered during the inference process of entity alignment.
- NMN (2020) [41]: This work proposes a neighborhood matching network that can handle the structural variability between KGs. The network utilizes similarity estimation between entities to capture both the topological structure and the difference in neighborhoods.
- CEA (2020) [46]: This work introduces a collective framework for formulating entity alignment (EA) as a standard stable matching problem. This framework is solved using the deferred acceptance algorithm.
- DAT (2020) [48]: This work proposes a degree-aware co-attention network that integrates semantic and structural features to enhance the performance of long-tail entities.
- DGMC (2020) [8]: This work introduces a two-stage neural architecture for acquiring and refining structural correspondences between graphs.
- AttrGNN (2020) [19]: Besides structural features, this research suggests utilizing an attributed value encoder and dividing the knowledge graph (KG) into subgraphs to model diverse types of attribute triples for alignment.
- RNM (2021) [53]: This work puts forward a relation-aware neighborhood matching model for entity alignment.
- CEAFF (2021) [47]: As an extension to CEA, this research suggests an adaptive feature fusion strategy to incorporate various features and a reinforcement learning-based model for conducting collective alignment.

We choose these baselines since they are the most recent and also the best performant approaches. Indeed, the majority of the baselines are embedding-based methods since most of the EA approaches merely focus on the embedding learning stage. There are only a limited number of methods that focus on the alignment inference stage, such as CEA, GM-EHD-JEA, and CEAFF. To ensure a fair comparison, we executed the source codes of the baseline methods in our experimental setup and presented the results obtained in comparison with the corresponding results reported in the original papers, despite the possibility of differences between the two. We have highlighted the top-performing results in each table by marking them in bold.

5.7 Results

We aim to answer the following research questions by conducting relevant experiments:

1. Can LIME effectively cope with large-scale datasets? (Sect. 5.7.1)
2. Can LIME outperform state-of-the-art solutions on datasets in normal scales? (Sect. 5.7.2)
3. What influence does the partition strategies have on the alignment process? (Sect. 5.7.3)
4. Is reciprocal alignment inference more effective than the frequently used CSLS metric? Could more insights into the reciprocal modeling process be provided? (Sect. 5.7.4)
5. Is the progressive blocking process sensitive to the hyper-parameters? How to set the hyper-parameters? (Sect. 5.7.5)

5.7.1 Evaluation on Large-Scale Dataset

Settings To address RQ1, we experimented on FB_DBP_2M. All of the state-of-the-art approaches *cannot* be directly implemented on this dataset due to the huge computation cost. Hence, we utilized the SBP algorithm to partition KGs and used CPS and I-SBP for comparison. We used GCN and RREA as the entity representation learning models. Regarding the alignment inference stage, we compared our proposed reciprocal alignment inference strategy RInf and its variant methods RInf-wr, RInf-pb, with the direct alignment strategy DInf. For the comprehensiveness of evaluation, we also conducted the experiments on the medium-sized dataset DWY100K. The results are presented in Table 5.2.

Overall Results According to Table 5.2, the best alignment performance on FB_DBP_2M and DWY100K$_{DBP-WD}$ is achieved by the combination of I-SBP, RREA,

Table 5.2 Evaluation results of variants of LIME on large-scale datasets

Partition	Embedding	Inference	FB_DBP_2M			DWY100K$_{DBP-WD}$			DWY100K$_{DBP-YG}$		
			Hits@1	Time	Mem	Hits@1	Time	Mem	Hits@1	Time	Mem
CPS	GCN	DInf	4.8	**4,900**	**13.3**	39.0	**66**	**5.4**	45.9	**64**	**5.4**
CPS	GCN	RInf	–	–	–	49.7	1,140	150.3	55.0	1,135	150.0
CPS	GCN	RInf-pb	5.2	5,761	18.6	47.4	120	6.7	53.1	143	8.4
CPS	GCN	RInf-wr	5.4	4,915	17.2	46.5	67	**5.4**	52.6	66	**5.4**
SBP	GCN	DInf	7.0	8,672	14.5	43.8	106	5.7	50.5	118	5.7
SBP	GCN	RInf	–	–	–	55.4	1,311	151.0	61.3	1,350	151.3
SBP	GCN	RInf-pb	7.4	12,229	20.1	53.1	180	8.2	58.9	279	13.0
SBP	GCN	RInf-wr	6.3	8,710	18.3	49.8	108	5.7	47.0	120	5.7
I-SBP	GCN	DInf	9.4	29,793	35.0	55.2	603	6.5	66.3	680	6.5
I-SBP	GCN	RInf	–	–	–	73.1	5,458	155.8	81.8	5,337	156.3
I-SBP	GCN	RInf-pb	17.7	39,596	35.7	79.4	1,071	9.6	86.3	1,434	13.1
I-SBP	GCN	RInf-wr	9.9	30,276	35.0	64.1	636	6.5	65.4	673	6.5
CPS	RREA	DInf	8.8	7,450	27.1	60.1	198	6.7	66.7	203	6.8
CPS	RREA	RInf	–	–	–	68.8	1,411	155.8	72.6	1,326	150.0
CPS	RREA	RInf-pb	9.4	7,642	27.1	67.2	240	7.2	71.1	250	8.6
CPS	RREA	RInf-wr	9.7	7,465	27.1	66.7	210	6.7	71.3	213	6.8
SBP	RREA	DInf	13.5	12,748	30.7	67.5	268	7.2	74.6	318	8.0
SBP	RREA	RInf	–	–	–	76.9	1,438	150.0	81.2	1,450	156.0
SBP	RREA	RInf-pb	14.4	14,031	30.7	74.7	318	7.7	79.1	407	11.0
SBP	RREA	RInf-wr	12.8	12,785	30.7	71.0	269	7.3	62.6	318	8.0
I-SBP	RREA	DInf	21.4	43,972	49.6	80.0	1,060	8.0	85.6	1,290	8.7
I-SBP	RREA	RInf	–	–	–	**91.4**	5,850	158.0	93.1	6,316	156.0
I-SBP	RREA	RInf-pb	**39.7**	54,643	49.6	91.2	1,638	8.7	**93.3**	2,059	14.1
I-SBP	RREA	RInf-wr	21.3	46,098	49.6	84.0	1,368	8.3	70.5	1,530	9.0

and RInf-pb. However, replacing RInf-pb with RInf in this combination leads to the highest Hits@1 on $DWY100K_{DBP-WD}$. In terms of efficiency, the combination of CPS, GCN, and DInf is the fastest across all three KG pairs. Additionally, the alignment results on $DWY100K$ are much higher, while the memory and time costs are lower than those on FB_DBP_2M, demonstrating that our newly constructed large-scale EA dataset presents a significant challenge to EA solutions.

Partition Strategies In terms of partition strategies, it is clear that I-SBP consistently achieves the best alignment results, regardless of the choice of embedding and inference models. Moreover, using SBP results in better alignment performance than using CPS, highlighting the effectiveness of leveraging bidirectional information for KG partitioning. However, SBP is more time- and memory-intensive than CPS as it requires bidirectional partitions. I-SBP further increases the time cost to a significantly higher level, at least three times that of SBP. This excessive time cost is due to the iterative re-partitioning process. Additionally, I-SBP consumes more memory space than other partition strategies.

Alignment Inference Strategies Initially, we compare our proposed alignment inference strategies with the direct alignment inference approach. The results presented in Table 5.2 indicate that our proposed reciprocal inference strategy RInf outperforms the commonly used direct alignment inference DInf by a significant margin on $DWY100K$. On the FB_DBP_2M dataset, although RInf cannot work due to the high memory cost, its approximation strategies still attain better results than DInf. Particularly, compared to DInf, RInf-pb only requires more time and memory space within a reasonable range while consistently achieving superior alignment results across all datasets under different combinations of partition and embedding learning models. It is especially effective in the iterative partition setting. On the other hand, RInf-wr incurs slightly higher time and memory costs than DInf and achieves better results than DInf when using CPS while performing worse than DInf under bidirectional partitioning on $DWY100K_{DBP-YG}$ and FB_DBP_2M. This can be attributed to the fact that directly aggregating the preference scores can result in the loss of information, as the preference scores are typically very close. This is also discussed in Sect. 5.4.3.

In the next step, we compare RInf with its variants. On $DWY100K$, applying the blocking strategy reduces the Hits@1 performance of RInf by 2–5%, with the exception of cases where I-SBP is used. This is because the blocking process cannot guarantee that equivalent entities are placed in the same block. Nonetheless, RInf-pb reduces the memory cost by over 90% and the time cost by over 70%. This validates that our progressive blocking strategy can significantly increase the efficiency of the reciprocal modeling process at the cost of a slight performance drop. Although applying the blocking strategy reduces the Hits@1 performance of LIME, the results are still significantly higher than DInf. When using the iterative partition strategy, it can be observed that using RInf-pb achieves comparable or even better Hits@1 performance than using RInf. This is because the progressive blocking process reduces the search space and can generate more confident pairs, which could lead to increasingly better partition and alignment results.

Regarding the no-ranking variant RInf-wr, even though its time and memory costs are small (close to DInf), its alignment performance is significantly lower than RInf across all settings. This confirms that the ranking process is crucial in the preference aggregation process, as discussed in Sect. 5.4.3.

Representation Learning Models Regarding the entity structural embedding learning methods, the more advanced model RREA achieves better results than the baseline model GCN with various partition and inference strategies, demonstrating the importance of modeling KG structure information for overall alignment performance. This also confirms that our proposal is independent of the embedding learning model and can consistently improve alignment results.

For further details on the design of each component and more experiments and discussions, please refer to the following subsections.

5.7.2 Comparison with State-of-the-Art Methods

In this subsection, we answer RQ2.

Settings In the previous section, we demonstrated that LIME can effectively handle large-scale EA datasets. However, since state-of-the-art methods cannot handle the FB_DBP_2M dataset, we conducted further experiments on popular medium-sized and small datasets to validate the effectiveness of our proposal. Given that these datasets are relatively smaller, we did not use our proposed partition strategies in LIME, as discussed in Sect. 5.3.5. Therefore, we evaluated the effectiveness of our proposed reciprocal alignment inference strategy and its variants, using the RREA model as the representation learning module in LIME. We denoted using the no-ranking and progressive blocking variants as LIME-wr and LIME-pb, respectively.

We presented the results of methods that only utilize KG structure to learn entity embeddings for alignment in Table 5.3 and the results of methods that use additional information in Table 5.4. Additionally, we demonstrated that LIME can be applied to other representation learning models, and the results are reported in Table 5.5. We also provided a comparison of efficiency in Fig. 5.6.

Comparison of Alignment Performance We can observe from Tables 5.3 and 5.4 that LIME achieves the best alignment performance in both categories, and the performance of LIME-wr and LIME-pb also surpasses that of the baseline models, validating the effectiveness of the reciprocal inference strategy and its variant strategies. Notably, LIME adopts RREA as the representation learning component, which has already attained the highest Hits@1 among existing methods. In order to further validate that LIME is a generic framework that can be used to improve the alignment performance of any representation learning-based EA method, we removed the RREA model and applied LIME to other models. We reported the corresponding results in Table 5.5. Specifically, we selected a representative approach from each group, namely, RSNs and RDGCN, and reported the results on DBP15K

Table 5.3 Hits@1 results of methods merely using structural information

Methods	DBP15K			SRPRS				DWY100K	
	ZH-EN	JA-EN	FR-EN	EN-FR	EN-DE	DBP-WD	DBP-YG	DBP-WD	DBP-YG
MTransE	20.9	25.0	24.7	21.3	10.7	18.8	19.6	23.8	22.7
KECG	47.7	49.2	48.5	29.8	44.4	32.3	35.0	63.1	71.9
MuGNN	47.0	48.3	49.1	13.1	24.5	15.1	17.5	60.4	73.9
RSNs	58.0	57.4	61.2	35.0	48.4	39.1	39.3	49.7	61.0
AliNet	55.5	55.3	56.2	28.7	43.5	33.7	35.3	60.4	71.5
MRAEA	65.8	65.9	67.4	40.3	54.3	45.6	46.9	72.3	83.4
TransEdge	59.4	55.1	55.8	35.0	50.7	39.4	38.8	66.6	75.6
MMEA*	68.1	65.5	67.7	–	–	–	–	–	–
HyperKA*	57.2	56.4	59.7	–	–	–	–	–	–
LDSD*	–	–	–	41.1	53.9	46.8	46.5	–	–
SSP	63.9	63.9	66.1	37.2	52.1	41.5	41.4	70.7	79.4
RREA	71.6	70.8	73.2	43.0	57.3	48.8	49.2	77.0	85.0
LIME	**75.6**	**75.3**	**78.6**	**45.5**	**58.6**	**51.2**	**50.6**	**81.6**	**88.0**
LIME-wr	73.6	73.1	75.6	44.4	58.1	50.2	50.2	78.7	86.5
LIME-pb	72.7	72.5	75.6	43.6	57.5	49.4	49.7	78.6	86.2

For some methods, we do not have access to their source codes and our implemented results are much worse than those reported in the original papers. In this case, we directly adopt the results from the papers (which might be missing on some datasets) and mark these methods with * in the tables

Table 5.4 Hits@1 results of methods using additional information

Methods	DBP15K			SRPRS				DWY100K	
	ZH-EN	JA-EN	FR-EN	EN-FR	EN-DE	DBP-WD	DBP-YG	DBP-WD	DBP-YG
JAPE	41.4	36.5	31.8	24.1	26.8	21.2	19.3	33.9	21.6
GCN-Align	43.4	42.7	41.1	29.6	42.8	32.7	34.7	51.3	59.6
RDGCN	69.7	76.3	87.3	67.2	77.9	97.4	99.0	–	–
HGCN	70.8	75.8	88.8	67.0	76.3	98.9	99.1	98.4	99.2
GM-EHD-JEA*	73.6	79.2	92.4	–	–	–	–	–	–
NMN	73.3	78.5	90.2	51.1	66.9	98.9	99.5	98.1	96.0
CEA	78.7	86.3	97.2	96.2	97.1	99.8	99.9	**100.0**	**100.0**
DAT	78.8	83.5	92.2	75.8	87.6	92.6	94.0	99.2	96.5
DGMC*	80.1	84.8	93.3	–	–	–	–	–	–
AttrGNN*	79.6	78.3	91.9	–	–	–	–	96.1	99.9
RNM	83.4	85.0	92.9	52.5	70.5	87.6	90.5	97.9	95.4
CEAFF	81.1	86.8	97.2	96.8	98.1	**100.0**	**100.0**	**100.0**	**100.0**
LIME	**87.4**	**90.9**	**97.8**	**97.5**	**98.6**	**100.0**	**100.0**	**100.0**	**100.0**
LIME-wr	86.5	90.3	97.1	97.0	98.2	**100.0**	**100.0**	**100.0**	**100.0**
LIME-pb	85.0	89.3	97.1	96.9	98.1	**100.0**	**100.0**	**100.0**	**100.0**

Table 5.5 Hits@1 results of applying LIME to other methods on DBP15K and SRPRS

Methods	DBP15K			SRPRS	
	ZH-EN	JA-EN	FR-EN	EN-FR	EN-DE
RSNs	58.0	57.4	61.2	35.0	48.4
+LIME	**63.6**	**62.7**	**68.4**	**38.8**	**53.0**
RDGCN	69.7	76.3	87.3	67.2	77.9
+LIME	**77.0**	**82.8**	**92.3**	**71.8**	**83.2**

and SRPRS in Table 5.5. Results on other datasets are omitted due to space limitations. The results in Table 5.5 verify that applying LIME leads to much better alignment performance than the direct alignment inference strategy, regardless of the approaches or datasets. This further demonstrates the effectiveness and generality of the LIME framework.

Additionally, we can observe several trends from the tables: (1) the results on DWY100K are higher than those on DBP15K and SRPRS, since the KGs in DWY100K are denser, which can provide more structural information for alignment. In comparison, the results on SRPRS are the worst among the three as its KG structure is the sparsest. This reveals that the density of the KG structure is crucial to the alignment of entities; and (2) overall, compared with methods that only use structural information, the methods that incorporate additional features achieve much better alignment performance. On the mono-lingual datasets, some solutions even achieve ground-truth results, showcasing the benefits of incorporating other useful features.

Usage of Partition Strategies In Sect. 5.3.5, we discussed that when dealing with small- or medium-sized datasets, it may not be worth using the partition strategy since partitioning may not significantly reduce computational costs, while it may decrease the alignment accuracy. We empirically validate this point by comparing the results of LIME in Table 5.3 with the results in Table 5.2. Specifically, we can see from Table 5.2 that the Hits@1 of SBP +RREA +RInf[6] on DWY100K$_{DBP-WD}$ is 76.9%, while this figure is 81.6% for LIME (equivalent to RREA +RInf) in Table 5.3, demonstrating that the partition process indeed harms the alignment accuracy. Furthermore, in terms of time cost, they are of the same order of magnitude (thousands of seconds), despite the fact that using the partition strategy would be faster.

Comparison of the Efficiency We compare LIME with state-of-the-art approaches in terms of the efficiency and show the results in Fig. 5.6.[7] The study demonstrates

---

[6] Note that we do not compare with I-SBP, since it selects confident EA pairs to augment training data, which improves both the partition and the representation learning process. It corresponds to the semi-supervised setting in previous EA works [22, 34], which is usually not compared with the methods without the semi-supervised setting (e.g., LIME in this work) for fairness.

[7] We do not include an evaluation of the efficiency of methods that use extra information because processing this information is complex and it is challenging to provide an unbiased evaluation.

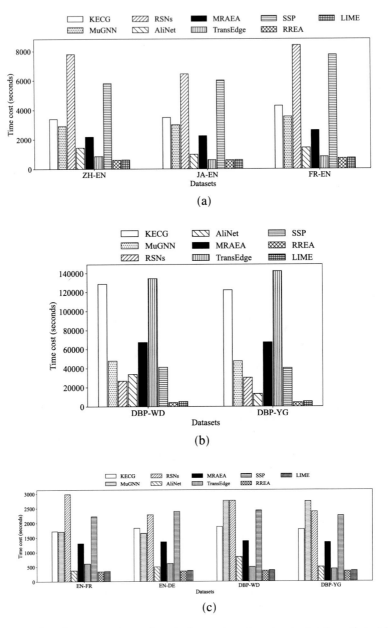

Fig. 5.6 Running time comparison of methods merely using structural information. (**a**) On DBP15K. (**b**) On DWY100K. (**c**) On SRPRS

that LIME is effective on all datasets, primarily because the representation learning model RREA is highly efficient. However, LIME does require slightly more running time to significantly enhance the alignment performance of RREA. It is also worth noting that the time cost is generally higher on larger datasets (such as DWY100K compared to DBP15K) and on denser datasets (such as DBP15K compared to SRPRS).

5.7.3   Experiments and Analyses on Partitioning

In this section, we seek to answer RQ3. By examining Table 5.2, we can conclude that I-SBP and SBP outperform CPS in generating precise alignment outcomes, albeit at the expense of greater time and memory usage. This section presents additional experiments aimed at assessing the efficacy of these partition methods.

Influence of Partition Strategies on Alignment Links   Our initial goal is to evaluate the percentage of preserved alignment links following the partitioning process. This is a critical aspect, as the optimal partition strategy should place equivalent entity pairs in the same subgraph pair, thereby enabling accurate alignment in subsequent stages. The ability of the partition strategy to group equivalent entity pairs together determines the maximum achievable alignment accuracy, as discussed in Sect. 5.3.2. As a result, we present the percentage of preserved gold alignment links following partitioning in Table 5.6.

It reads from Table 5.6 that CPS destroys over 10% of the links on DWY100K and more than half of the links on FB_DBP_2M. This indicates that the partition process itself significantly reduces the maximum achievable alignment accuracy, which is undesirable. In comparison, adopting SBP can retain 67.5% of the links on FB_DBP_2M, which increases the result of CPS by over 50%. Moreover, I-SBP produces a remarkable improvement, preserving 80% of the links in FB_DBP_2M and almost all links in DWY100K. This demonstrates that iterative partitioning can effectively optimize the partition process and prevent equivalent entities from being placed into different subgraph pairs. Nevertheless, as shown in Table 5.2, the time

Table 5.6 The percentage of gold alignment links preserved after partitioning

Dataset	Strategy	Overall	Train	Test
DWY100K$_{DBP-WD}$	CPS	89.5%	92.7%	88.1%
	SBP	97.6%	99.0%	97.0%
	I-SBP	99.6%	100.0%	99.4%
DWY100K$_{DBP-YG}$	CPS	89.3%	92.7%	87.9%
	SBP	96.6%	98.0%	96.1%
	I-SBP	99.8%	100.0%	99.7%
FB_DBP_2M	CPS	43.7%	57.1%	38.0%
	SBP	67.5%	81.6%	61.4%
	I-SBP	81.0%	93.5%	75.6%

Table 5.7 The percentage of
preserved links, the time cost,
and the number of entities in
the largest subgraph pair of
CPS and SBP given
different k on FB_DBP_2M

Strategy	k	Presv. Links	Time	#Ents
CPS	50	45.2%	3,051	96,357
	75	43.7%	3,976	64,237
	100	38.1%	4,812	48,185
SBP	50	68.4%	5,834	181,205
	75	67.5%	7,335	120,367
	100	59.2%	9,016	92,935

cost of SBP is almost double that of CPS, while I-SBP requires significantly more time depending on the number of iterations.

Influence of the Number of Subgraph Pairs k   Our next step is to analyze the impact of the number of subgraph pairs k on the partition process. To be specific, Table 5.7 presents the percentage of preserved links, time cost, and the number of entities in the largest subgraph pair for CPS and SBP, with k set to 50, 75, and 100. Indeed, Table 5.7 indicates that as the number of subgraph pairs increases, the percentage of preserved links decreases, and the partition time cost increases for both CPS and SBP. However, increasing the number of subgraph pairs results in smaller subgraphs, which can be beneficial for structural representation learning strategies due to their scalability limitations.

5.7.4   Experiments and Analyses on Reciprocal Inference

In this subsection, we address RQ4.

Comparison with the CSLS Metric   In Sect. 5.4, we mentioned the CSLS metric, which was introduced to address the hubness problem in nearest neighbor search and may have a similar effect as the reciprocal alignment inference strategy. We thus replaced the reciprocal inference approach in LIME with the CSLS metric (with a hyper-parameter n set to 1, 5, or 10) and evaluated the corresponding Hits@1 results, which are presented in Fig. 5.7. It is worth noting that all other settings were kept the same.

The results presented in Fig. 5.7 demonstrate that LIME consistently outperforms the CSLS metric on all datasets. This confirms that our reciprocal alignment inference strategy can more effectively model and integrate entity preferences, leading to more accurate alignment results compared to the CSLS metric (as discussed in Sect. 5.4). Additionally, we observe that the performance of the CSLS metric deteriorates as the hyper-parameter n increases.

Deeper Insights into the Preference Modeling and Aggregation   It is worth noting that in cases where the entity representation learning model has poor performance on EA (i.e., the model outputs a homogeneous probability distribution

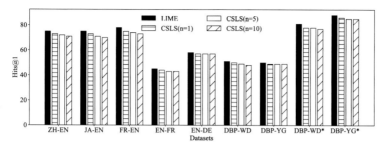

Fig. 5.7 Comparison of the reciprocal inference in LIME and the CSLS metric. DBP-WD* and DBP-YG* refer to the KG pairs in DWY100K. The results on FB_DBP_2M are omitted due to the excessive time and memory costs required by reciprocal inference in LIME and the CSLS metric

of entity embeddings), the preference matrix can have many ties, which may impede the effectiveness of the reciprocal modeling approach. Therefore, we aim to (1) analyze the likelihood of ties occurring in the preference matrix and (2) empirically demonstrate that our proposed inference strategy can still improve the performance of a low-performing entity representation learning model even in the presence of ties.

Take SRPRS$_{EN-FR}$, for example. We conducted an analysis of the preference matrices for RREA and a low-performing model RSNs, which have dimensions of 10,500*10,500. On average, ties occur 8.72 times in each row or column of the RREA preference matrix, which is not a frequent occurrence. For the low-performing RSNs model, this figure increases to 12.82. This suggests that the quality of entity representations can influence the frequency of ties during preference aggregation, but the effect is not significant. Furthermore, despite the presence of ties in the ranking matrices, applying our reciprocal inference strategy improves the Hits@1 of RSNs by 10.9% as shown in Table 5.5. This demonstrates that the reciprocal modeling approach can still benefit a low-performing entity representation learning method.

5.7.5   Experiments and Analyses on Progressive Blocking

In this subsection, we proceed to answer RQ5. First, we analyze the impact of the hyper-parameter θ on the alignment performance and efficiency. Next, we discuss the parameter settings of the progressive blocking process.

Analysis of θ   As mentioned in Sect. 5.5.2, setting θ to a small value will retain the majority of connections, resulting in most entities being placed in the same block. On the other hand, setting θ to a large value will remove many connections and separate entities into different isolated blocks. To empirically verify this claim, we conducted an experiment on the DBP15K$_{ZH-EN}$ dataset, varying the value of θ. We reported the total number of blocks (#Total), the size of the largest block

Table 5.8 Analysis of the hyper-parameter θ in progressive blocking on DBP15K$_{\text{ZH-EN}}$

θ	#Total	#MaxSize	#Iso	Perc.	H@1	H@1*
0.75	11,447	81	8,225	71.9%	48.1	67.9
0.70	9,345	161	6,331	67.7%	53.9	67.8
0.65	7,245	521	4,757	65.7%	58.1	67.5
0.60	5,034	1,917	3,463	68.8%	60.9	66.4
0.55	3,128	14,054	2,424	77.5%	62.7	66.0
0.50	1,825	17,910	1,569	86.0%	63.7	65.3
0.45	949	19,699	847	89.3%	64.6	65.2
0.40	437	20,421	407	93.1%	65.0	65.1

(#MaxSize), the number of blocks that only contain one entity (which we refer to as isolated blocks, #Iso), the percentage of isolated blocks (Perc.), the aggregated Hits@1 results of performing the alignment within each block (H@1), and the aggregated Hits@1 results of performing the alignment within each block and the aggregated isolated blocks (H@1*), in Table 5.8.

The results in Table 5.8 show that setting θ to a large value, specifically 0.75, results in the removal of most pairwise connections, leading to over 10,000 blocks, of which 71.9% were isolated blocks. Also, the Hits@1 result is very low (at 48.1%). Aggregating the 8,225 isolated blocks and considering the alignment performance within this aggregated block, the Hits@1 result increased to 67.9%. However, this aggregated block contains over 8,000 entities and still requires significant memory space. In contrast, setting θ to a small value, specifically 0.4, results in the majority of entities (over 20,000) being placed in the same block, which does not achieve the objective of reducing memory space.

Therefore, in a progressive blocking setting, the value of θ in the first round is typically set to a larger value. Although this may result in a larger size of the aggregated isolated block, the subsequent rounds with lower θ values further process the aggregated block.

Analysis of the Progressive Blocking In this work, we conduct three rounds of progressive blocking and directly set Θ to the 50th percentile (median), 25th percentile (the first quartile), and 1st percentile of the set of the largest similarity scores of all source entities, respectively. In this study, our goal is to investigate the impact of the values of θ and the number of rounds of progressive blocking on the alignment performance and memory consumption. To be more specific, we keep two threshold values constant and vary the value of the other threshold. Then, we report the Hits@1 and memory size in Fig. 5.8a, b, and c. Moreover, we perform progressive blocking for 0 to 4 rounds and present the Hits@1 and memory size in Fig. 5.8d.

As shown in Fig. 5.8a, the value of the initial threshold has an impact on the final Hits@1 result and memory cost. Setting the initial threshold to a relatively small value may produce more accurate alignment results, but it also comes with a high

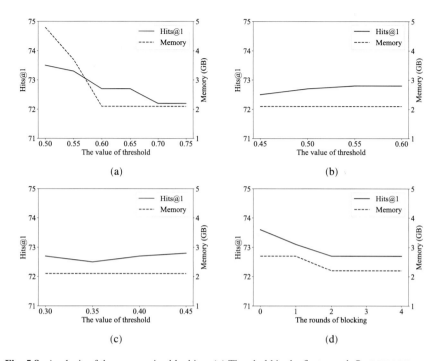

Fig. 5.8 Analysis of the progressive blocking. (**a**) Threshold in the first round. On DBP15K$_{ZH-EN}$. (**b**) Threshold in the second round. On DBP15K$_{ZH-EN}$. (**c**) Threshold in the third round. On DBP15K$_{ZH-EN}$. (**d**) Rounds of progressive blocking. On DBP15K$_{ZH-EN}$

memory cost since most entities are still connected and placed in the same block. On the other hand, a larger threshold can reduce the memory cost, but it also leads to a lower alignment performance.

Figures 5.8b and c demonstrate that the values of the thresholds in the second and third round do not have a significant impact on the memory cost, while they only have a small influence on the alignment performance. Furthermore, Fig. 5.8d indicates that the Hits@1 performance and memory cost drop in the first few rounds and remain relatively stable with more rounds of blocking. Therefore, conducting progressive blocking for a few rounds is sufficient.

Threshold Setting in Practice Based on the analysis, we can identify two crucial factors when setting the threshold schedule: (1) the threshold values should be gradually decreased; and (2) the initial threshold value should be selected carefully, possibly with the guidance of statistical information regarding the similarity scores. Therefore, our proposed strategy for scheduling the threshold is a feasible option in practice, and it can be adjusted based on the statistical information available.

5.8   Related Work

We will provide a brief overview of the studies that have addressed the scalability problem in EA. The experimental paper on EA [50] indicates that even the state-of-the-art EA methods still suffer from poor scalability. While simpler models such as GCN-Align [38] and ITransE [51] are faster, they tend to have poorer effectiveness [50]. In contrast, more effective models typically have complex architectures and are inefficient.

There have been several studies on relevant tasks that propose strategies for handling large-scale data. For instance, Flamino et al. [9] approach the alignment of entities in large-scale networks by clustering nodes using network-specific features. However, these features are not present in KGs, and the structure of KGs is more intricate than networks. Zhuang et al. [54] suggest partitioning entities from various knowledge bases into smaller blocks using predicates in the triples. Nonetheless, aligning predicates in different knowledge bases is already a challenging task, and the source codes of these methods are not available. Therefore, their implemented programs cannot be applied to the EA task. Zhang et al. [49] address the problem of linking large-scale heterogeneous entity graphs. However, the entity graph only includes entities in a few types, such as paper, author, and venue, and the relation types are also limited, which is very different from KGs. Thus, their proposed method, which depends on the characteristics of entity graphs, cannot be used for EA.

Several recent works have focused on addressing the efficiency issue in EA. Mao et al. [20] identify over-complex graph encoders and inefficient negative sampling strategies as the primary causes of poor efficiency in EA. They propose a novel KG encoder, Dual Attention Matching Network, to reduce computational complexity. However, their work focuses only on the representation learning stage and is evaluated on a medium-sized dataset, DWY100K. GM-EHD-JEA [42] formulates EA as a task assignment problem and proposes to solve it using the Hungarian algorithm. However, the Hungarian algorithm cannot be directly applied to EA due to its extra-large computation time. Therefore, they propose a space separation strategy to reduce the search space so that the Hungarian algorithm can work properly. This method is similar to our blocking strategy without the progressive procedure. However, we improve the performance by aggregating isolated blocks, and our progressive blocking process can further enhance efficiency.

Another recent work proposes a unidirectional strategy, CPS, to partition large-scale KGs and uses name information to improve alignment performance [10]. However, in general, the scalability issue in EA remains a critical and underexplored problem. It is worth noting that ER can be regarded as the general version of the EA task [50]. There have been several studies on improving the efficiency and scalability of ER, and we refer readers to the survey paper [7]. Our blocking strategy is inspired by these relevant works on ER.

5.9 Conclusion

In this chapter, we have highlighted the scalability issue in state-of-the-art EA approaches and proposed an effective solution, LIME, to address EA at scale. The LIME approach initially uses graph partition strategies that focus on seeds to divide large-scale KGs into smaller pairs. Then, LIME employs a novel reciprocal alignment inference strategy within each subgraph pair to generate alignment results based on the entity representations learned by existing embedding learning models. To enhance the scalability of reciprocal alignment inference, LIME suggests two variant strategies that can reduce computational costs, albeit with a slight decrease in performance. The experimental evaluations conducted on a novel large-scale EA dataset reveal that LIME can successfully address EA on a large scale. Besides, the empirical results on the popular EA datasets also validate the superiority of LIME and show that it can be applied to existing methods to improve their performance.

Appendix

Correctness Analysis

In Sect. 5.4.4, we examine the performance of reciprocal alignment inference and the baseline model (direct inference) under various conditions.

Case 1 For the ground-truth entity pair (u, v):

$$u = \arg\max\{sim(\boldsymbol{v}, \boldsymbol{u}'), u' \in \mathcal{E}_s\};$$
$$v = \arg\max\{sim(\boldsymbol{u}, \boldsymbol{v}'), v' \in \mathcal{E}_t\},$$

As previously discussed in Sect. 5.4.4, when an accurate similarity signal is provided by the representation learning process, both reciprocal alignment inference and direct alignment inference can produce the correct alignment.

Case 2 For the ground-truth entity pair (u, v):

$$u = \arg\max\{sim(\boldsymbol{v}, \boldsymbol{u}'), u' \in \mathcal{E}_s\};$$
$$v^* = \arg\max\{sim(\boldsymbol{u}, \boldsymbol{v}'), v' \in \mathcal{E}_t\}, \quad v^* \neq v,$$

In this case, the direct alignment inference *cannot* generate the correct answer, since it only considers u's preference and would generate (u, v^*) as the answer. In comparison, reciprocal alignment inference does not necessarily generate the correct answer. This is because we only know $p_{u,v} = 1$, $p_{v,u} < 1$. Given any target entity v', we can derive $p_{u,v'} \leq 1$, $p_{v',u} \leq 1$. Thus, $r_{u,v} \leq r_{u,v'}$, while we cannot compare

$r_{v,u}$ and $r_{v',u}$. Therefore, we also cannot compare $p_{u \leftrightarrow v} = (r_{u,v} + r_{v,u})/2$ with $p_{u \leftrightarrow v'} = (r_{u,v'} + r_{v',u})/2$ as the exact values are unknown.

Case 3 For the ground-truth entity pair (u, v):

$$u^* = \arg\max\{sim(\boldsymbol{v}, \boldsymbol{u}'), u' \in \mathcal{E}_s\}, \quad u^* \neq u;$$

$$v = \arg\max\{sim(\boldsymbol{u}, \boldsymbol{v}'), v' \in \mathcal{E}_t\},$$

In this case, the direct alignment inference can generate the correct answer since it only considers u's preference. In comparison, reciprocal alignment inference does not necessarily generate the correct answer. This is because we only know $p_{u,v} < 1$, $p_{v,u} = 1$. Given any target entity v', we can derive $p_{u,v'} \leq 1$, $p_{v',u} < 1$. Thus, $r_{v,u} < r_{v',u}$, while we cannot compare $r_{u,v}$ and $r_{u,v'}$. Therefore, we also cannot compare $p_{u \leftrightarrow v} = (r_{u,v} + r_{v,u})/2$ with $p_{u \leftrightarrow v'} = (r_{u,v'} + r_{v',u})/2$ as the exact values are unknown.

Case 4 For the ground-truth entity pair (u, v):

$$u^* = \arg\max\{sim(\boldsymbol{v}, \boldsymbol{u}'), u' \in \mathcal{E}_s\}, \quad u^* \neq u;$$

$$v^* = \arg\max\{sim(\boldsymbol{u}, \boldsymbol{v}'), v' \in \mathcal{E}_t\}, \quad v^* \neq v,$$

In this case, the direct alignment inference *cannot* generate the correct answer, since it only considers u's preference and would generate (u, v^*) as the answer. In comparison, reciprocal alignment inference does not necessarily generate the correct answer. This is because we only know $p_{u,v} < 1$, $p_{v,u} < 1$. Given any target entity v', we can derive $p_{u,v'} \leq 1$, $p_{v',u} \leq 1$. However, we cannot compare $r_{v,u}$ and $r_{v',u}$, or $r_{u,v}$ and $r_{u,v'}$. Therefore, we cannot compare $p_{u \leftrightarrow v} = (r_{u,v} + r_{v,u})/2$ with $p_{u \leftrightarrow v'} = (r_{u,v'} + r_{v',u})/2$ as the exact values are unknown.

To summarize, the direct alignment inference method can only provide correct results in Case 1 and Case 3, while our proposed reciprocal alignment inference strategy can generate correct answers in Case 1 and has the potential to produce correct results in other cases as well. Since the input similarity matrix is often not very accurate, as representation learning models may not fully capture the relatedness between entities, our proposed method is expected to perform better than direct alignment inference, as empirically demonstrated in our experiments.

References

1. S. Auer, C. Bizer, G. Kobilarov, J. Lehmann, R. Cyganiak, and Z. G. Ives. Dbpedia: A nucleus for a web of open data. In *ISWC*, pages 722–735, 2007.
2. A. Bordes, N. Usunier, A. García-Durán, J. Weston, and O. Yakhnenko. Translating embeddings for modeling multi-relational data. In *NIPS*, pages 2787–2795, 2013.

3. F. Bourse, M. Lelarge, and M. Vojnovic. Balanced graph edge partition. In *KDD*, pages 1456–1465. ACM, 2014.

4. Y. Cao, Z. Liu, C. Li, Z. Liu, J. Li, and T. Chua. Multi-channel graph neural network for entity alignment. In *ACL*, pages 1452–1461, 2019.

5. J. Chen, Z. Li, P. Zhao, A. Liu, L. Zhao, Z. Chen, and X. Zhang. Learning short-term differences and long-term dependencies for entity alignment. In *ISWC*, volume 12506, pages 92–109, 2020.

6. M. Chen, Y. Tian, M. Yang, and C. Zaniolo. Multilingual knowledge graph embeddings for cross-lingual knowledge alignment. In *IJCAI*, pages 1511–1517, 2017.

7. V. Christophides, V. Efthymiou, T. Palpanas, G. Papadakis, and K. Stefanidis. An overview of end-to-end entity resolution for big data. *ACM Comput. Surv.*, 53(6), 2020.

8. M. Fey, J. E. Lenssen, C. Morris, J. Masci, and N. M. Kriege. Deep graph matching consensus. In *ICLR*. OpenReview.net, 2020.

9. J. Flamino, C. Abriola, B. Zimmerman, Z. Li, and J. Douglas. Robust and scalable entity alignment in big data. In *IEEE Big Data*, pages 2526–2533, 2020.

10. C. Ge, X. Liu, L. Chen, B. Zheng, and Y. Gao. Largeea: Aligning entities for large-scale knowledge graphs. *CoRR*, abs/2108.05211, 2021.

11. L. Guo, Z. Sun, and W. Hu. Learning to exploit long-term relational dependencies in knowledge graphs. In *ICML*, pages 2505–2514, 2019.

12. G. Karypis and V. Kumar. A fast and high quality multilevel scheme for partitioning irregular graphs. *SIAM J. Sci. Comput.*, 20(1):359–392, 1998.

13. G. Karypis and V. Kumar. Metis: a software package for partitioning unstructured graphs. 1998.

14. G. Karypis and V. Kumar. Multilevel k-way partitioning scheme for irregular graphs. *J. Parallel Distributed Comput.*, 48(1):96–129, 1998.

15. T. N. Kipf and M. Welling. Semi-supervised classification with graph convolutional networks. In *ICLR*. OpenReview.net, 2017.

16. G. Lample, A. Conneau, M. Ranzato, L. Denoyer, and H. Jégou. Word translation without parallel data. In *ICLR*. OpenReview.net, 2018.

17. C. Li, Y. Cao, L. Hou, J. Shi, J. Li, and T. Chua. Semi-supervised entity alignment via joint knowledge embedding model and cross-graph model. In *EMNLP*, pages 2723–2732. Association for Computational Linguistics, 2019.

18. L. Li and T. Li. MEET: a generalized framework for reciprocal recommender systems. In *CIKM*, pages 35–44. ACM, 2012.

19. Z. Liu, Y. Cao, L. Pan, J. Li, and T. Chua. Exploring and evaluating attributes, values, and structures for entity alignment. In *EMNLP*, pages 6355–6364, 2020.

20. X. Mao, W. Wang, Y. Wu, and M. Lan. Boosting the speed of entity alignment 10 ×: Dual attention matching network with normalized hard sample mining. In *WWW*, pages 821–832, 2021.

21. X. Mao, W. Wang, H. Xu, M. Lan, and Y. Wu. MRAEA: an efficient and robust entity alignment approach for cross-lingual knowledge graph. In *WSDM*, pages 420–428. ACM, 2020.

22. X. Mao, W. Wang, H. Xu, Y. Wu, and M. Lan. Relational reflection entity alignment. In *CIKM*, pages 1095–1104, 2020.

23. J. Neve and I. Palomares. Aggregation strategies in user-to-user reciprocal recommender systems. In *SMC*, pages 4031–4036. IEEE, 2019.

24. H. Nie, X. Han, L. Sun, C. M. Wong, Q. Chen, S. Wu, and W. Zhang. Global structure and local semantics-preserved embeddings for entity alignment. In *IJCAI*, pages 3658–3664, 2020.

25. I. Palomares, C. Porcel, L. A. Pizzato, I. Guy, and E. Herrera-Viedma. Reciprocal recommender systems: Analysis of state-of-art literature, challenges and opportunities on social recommendation. *CoRR*, abs/2007.16120, 2020.

26. G. Papadakis, D. Skoutas, E. Thanos, and T. Palpanas. Blocking and filtering techniques for entity resolution: A survey. *ACM Comput. Surv.*, 53(2):31:1–31:42, 2020.

27. L. A. S. Pizzato, T. Rej, T. Chung, I. Koprinska, and J. Kay. RECON: a reciprocal recommender for online dating. In *RecSys*, pages 207–214. ACM, 2010.

28. X. Shi and Y. Xiao. Modeling multi-mapping relations for precise cross-lingual entity alignment. In *EMNLP*, pages 813–822. Association for Computational Linguistics, 2019.
29. J. Stoyanovich, B. Howe, and H. V. Jagadish. Responsible data management. *Proc. VLDB Endow.*, 13(12):3474–3488, 2020.
30. F. M. Suchanek, G. Kasneci, and G. Weikum. Yago: a core of semantic knowledge. In *WWW*, pages 697–706, 2007.
31. Z. Sun, M. Chen, W. Hu, C. Wang, J. Dai, and W. Zhang. Knowledge association with hyperbolic knowledge graph embeddings. In *EMNLP*, pages 5704–5716, 2020.
32. Z. Sun, W. Hu, and C. Li. Cross-lingual entity alignment via joint attribute-preserving embedding. In *ISWC*, pages 628–644, 2017.
33. Z. Sun, W. Hu, Q. Zhang, and Y. Qu. Bootstrapping entity alignment with knowledge graph embedding. In *IJCAI*, pages 4396–4402, 2018.
34. Z. Sun, J. Huang, W. Hu, M. Chen, L. Guo, and Y. Qu. Transedge: Translating relation-contextualized embeddings for knowledge graphs. In *ISWC*, pages 612–629, 2019.
35. Z. Sun, C. Wang, W. Hu, M. Chen, J. Dai, W. Zhang, and Y. Qu. Knowledge graph alignment network with gated multi-hop neighborhood aggregation. In *AAAI*, 2020.
36. X. Tang, J. Zhang, B. Chen, Y. Yang, H. Chen, and C. Li. BERT-INT: A bert-based interaction model for knowledge graph alignment. In *IJCAI*, pages 3174–3180, 2020.
37. D. Vrandecic and M. Krötzsch. Wikidata: a free collaborative knowledgebase. *Commun. ACM*, 57(10):78–85, 2014.
38. Z. Wang, Q. Lv, X. Lan, and Y. Zhang. Cross-lingual knowledge graph alignment via graph convolutional networks. In *EMNLP*, pages 349–357, 2018.
39. Y. Wu, X. Liu, Y. Feng, Z. Wang, R. Yan, and D. Zhao. Relation-aware entity alignment for heterogeneous knowledge graphs. In *IJCAI*, pages 5278–5284, 2019.
40. Y. Wu, X. Liu, Y. Feng, Z. Wang, and D. Zhao. Jointly learning entity and relation representations for entity alignment. In *EMNLP*, pages 240–249. Association for Computational Linguistics, 2019.
41. Y. Wu, X. Liu, Y. Feng, Z. Wang, and D. Zhao. Neighborhood matching network for entity alignment. In *ACL*, pages 6477–6487. Association for Computational Linguistics, 2020.
42. K. Xu, L. Song, Y. Feng, Y. Song, and D. Yu. Coordinated reasoning for cross-lingual knowledge graph alignment. In *AAAI*, pages 9354–9361. AAAI Press, 2020.
43. R. R. Yager and A. N. Rybalov. Uninorm aggregation operators. *Fuzzy Sets Syst.*, 80(1):111–120, 1996.
44. H. Yang, Y. Zou, P. Shi, W. Lu, J. Lin, and X. Sun. Aligning cross-lingual entities with multi-aspect information. In *EMNLP*, pages 4430–4440. Association for Computational Linguistics, 2019.
45. W. Zeng, X. Zhao, J. Tang, X. Li, M. Luo, and Q. Zheng. Towards entity alignment in the open world: An unsupervised approach. In *DASFAA*, volume 12681, pages 272–289, 2021.
46. W. Zeng, X. Zhao, J. Tang, and X. Lin. Collective entity alignment via adaptive features. In *ICDE*, pages 1870–1873. IEEE, 2020.
47. W. Zeng, X. Zhao, J. Tang, X. Lin, and P. Groth. Reinforcement learning based collective entity alignment with adaptive features. *ACM Transactions on Information Systems*, 39(3), 2021.
48. W. Zeng, X. Zhao, W. Wang, J. Tang, and Z. Tan. Degree-aware alignment for entities in tail. In *SIGIR*, pages 811–820. ACM, 2020.
49. F. Zhang, X. Liu, J. Tang, Y. Dong, P. Yao, J. Zhang, X. Gu, Y. Wang, B. Shao, R. Li, and K. Wang. OAG: toward linking large-scale heterogeneous entity graphs. In *SIGKDD*, pages 2585–2595, 2019.
50. X. Zhao, W. Zeng, J. Tang, W. Wang, and F. Suchanek. An experimental study of state-of-the-art entity alignment approaches. *IEEE Transactions on Knowledge and Data Engineering*, pages 1–1, 2020.
51. H. Zhu, R. Xie, Z. Liu, and M. Sun. Iterative entity alignment via joint knowledge embeddings. In *IJCAI*, pages 4258–4264, 2017.
52. Q. Zhu, X. Zhou, J. Wu, J. Tan, and L. Guo. Neighborhood-aware attentional representation for multilingual knowledge graphs. In *IJCAI*, pages 1943–1949, 2019.

53. Y. Zhu, H. Liu, Z. Wu, and Y. Du. Relation-aware neighborhood matching model for entity alignment. *AAAI*, pages 4749–4756, 2021.
54. Y. Zhuang, G. Li, Z. Zhong, and J. Feng. Hike: A hybrid human-machine method for entity alignment in large-scale knowledge bases. In *CIKM*, pages 1917–1926, 2017.

Chapter 6
Long-Tail Entity Alignment

Abstract Most entity alignment solutions currently rely on structural information, specifically KG embedding, to align entities. However, in real-life KGs, the majority of entities have a sparse neighborhood structure, while only a few entities are densely connected to others. These less-connected entities are referred to as *long-tail entities*, and this phenomenon limits the effectiveness of using structural information for entity alignment.

To address this issue, we propose an approach that incorporates *entity name* information, which is often overlooked but readily available. We amplify the weak structural information of long-tail entities with concatenated power mean word embeddings of their names during pre-alignment. To align entities, we introduce a novel complementary framework that combines both structural and name signals. It uses the entity's *degree* as a guide to fuse the two sources of information effectively and proposes a degree-aware *co-attention* network that dynamically adjusts the significance of features in a degree-aware manner. Finally, we propose using confident entity alignment results as anchors to complement original KGs with facts from their counterparts via *iterative* training during post-alignment. Experimental evaluations show the effectiveness of the proposed techniques.

6.1 Introduction

Many current approaches to entity alignment (EA) in knowledge graphs (KGs) heavily rely on the graph structure of KGs [4, 7, 10, 15, 18]. These approaches assume that equivalent entities have similar neighborhood structures. While these methods have achieved state-of-the-art performance on synthetic datasets extracted from large-scale KGs, as mentioned in [2, 15, 24], recent studies have shown that these *synthetic* datasets are much denser than real-life KGs. Furthermore, existing EA methods are *not* capable of yielding satisfactory results on datasets with real-life distributions, as discussed in [7].

A recent study, referenced as [7], has shown that nearly half of the entities in actual knowledge graphs have connections to less than three other entities, which are called *long-tail entities*. This results in the KG being a relatively sparse

© The Author(s) 2023
X. Zhao et al., *Entity Alignment*, Big Data Management,
https://doi.org/10.1007/978-981-99-4250-3_6

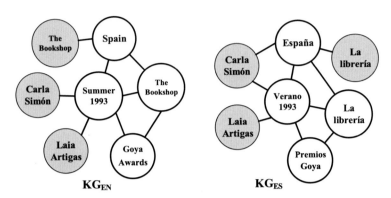

Fig. 6.1 An example of EA. Nodes in gray (resp. white) are long-tail (resp. popular) entities (relation names and other entities are omitted in the interest of space)

graph. This matches our perception that only a few entities in real-life KGs are frequently accessed and have rich connections and detailed attributes, while the majority remain under-explored and provide little structural information. This leads to existing EA methods that rely solely on structural information struggling to accurately align these entities, as demonstrated in the following example.

> *Example* In Fig. 6.1 is a partial English KG (KG$_{EN}$) and a partial Spanish KG (KG$_{ES}$) concerning the film Summer 1993. Note that the entities The Bookshop and La Librería in gray describe the original novel, while those in white depict the film.
>
> During aligning entities of high degrees, e,g., Spain and España, structural information is of great help; however, as to long-tail entities, e.g., Carla Simón in KG$_{EN}$, structural information may suggest Laia Artigas in KG$_{ES}$ as its match, since they have a single link to Summer 1993 and Verano 1993, respectively.

The example unveils the shortcoming of solely relying on structural information for EA, which renders existing EA methods sup-optimal, even infeasible for long-tail entities. Hence, we are motivated to revisit the key phases of EA pipeline and address the challenge of EA when structural information is insufficient.

In the *pre-alignment* phase, we are searching for extra signals that can improve EA, and we find that entity names can provide a source of valuable information. This type of information is commonly present in real-life entities, but previous research has not given it sufficient attention. For example, if we consider the long-tail entity Carla Simón in KG$_{EN}$, incorporating entity name information would be beneficial in finding the correct mapping, which is Carla Simón in KG$_{ES}$.

This shows that entity name information can provide a supplementary perspective to the commonly used structural information in EA.

Previous studies [19–21] have already used name embeddings, specifically averaged word embeddings, to populate initial feature matrices for learning *structural representation*. However, our approach is different in that we use entity names as an additional source of signal, in addition to structural information. We achieve this by encoding the names through concatenated power mean word embeddings [22].

During the *alignment* phase, we carefully merge the two signals mentioned earlier by considering the fact that the significance of structural and name information differs for entities with varying degrees. In the example above, aligning the long-tail entity Carla Simón in KG_{EN} relies more on entity name information than its limited neighboring structure. Conversely, for mapping popular entities such as the film La Librería, where ambiguous entity names are present (i.e., both the film La Librería and the novel La Librería share the same name), structure plays a more significant role. Generally speaking, we can assume that the importance of the entity name signal is higher (resp. lower) for entities with lower (resp. higher) degrees, while the opposite is true for the signal from neighboring structure. In order to accurately represent the nonlinear dynamics between the two signals, we develop a co-attention network that uses entity degrees as a guide to determine the weights of various signals. It is important to note that [10] introduced degrees as a way to address the bias in *structural embedding* methods, which tend to place entities with similar degrees in close proximity. However, our motivation is different in that we use degrees to calculate pairwise similarities instead of individual embeddings.

During the *post-alignment* phase, our proposal is to significantly improve the structural information of knowledge graphs by recursively examining and cross-referencing each other. While long-tail entities may lack structural information in their original knowledge graph (referred to as the "source KG"), the knowledge graph being aligned with (the "target KG") may have this information in a complementary manner. As an illustration, let's consider the entity Carla Simón. In KG_{EN}, there may be missing information such as the fact that Carla Simón is from España, which is present in KG_{ES}. By pairing the surrounding entities and leveraging information from the target KG, the source KG can potentially acquire this missing information and improve the alignment. Inspired by the beneficial impact of using rules to complete knowledge graphs [2], we propose an *iterative* training procedure that includes knowledge graph completion. In each round, we use confident entity alignment results as anchors to identify and add any missing relations, thereby enhancing the current knowledge graphs. As a result, these knowledge graphs become enriched, which in turn allows for the learning of better structural embeddings. Additionally, the matching signal can propagate to long-tail entities, which were previously difficult to align in a single shot but may now become easier to align as a result of this iterative process.

Contribution In short, the contribution of this chapter can be summarized as follows:

- We have observed a shortcoming in current EA methods regarding the alignment of long-tail entities, primarily because they heavily rely on structure. To overcome this limitation, we propose two solutions: (1) incorporating an additional signal from entity names through concatenated power mean word embeddings and (2) devising an efficient degree-aware co-attention mechanism to dynamically integrate the name and structural signals.
- Our proposal aims to decrease the number of long-tail entities by enhancing relational structure through KG completion, integrated into an iterative self-training approach. This is achieved by utilizing confident EA outcomes as anchors and using other KGs as references. Our strategy not only improves the performance of EA but also enhances the coverage of KGs.
- The techniques presented form a new framework called DAT. We conduct empirical evaluations of the implementation of DAT on both mono-lingual and cross-lingual EA tasks, comparing it to state-of-the-art methods. The results of our comparison and ablation analysis demonstrate the superiority of DAT.

Organization Section 6.2 overviews related work. In Sect. 6.3, we analyze the long-tail phenomenon in EA. DAT and its components are elaborated in Sect. 6.4. Section 6.5 introduces experimental settings, evaluation results, and detailed analysis, followed by conclusion in Sect. 6.6.

6.2 Related Work

Conventional EA Framework The advancements made by state-of-the-art methods can be analyzed based on a phased pipeline. Firstly, for the pre-alignment phase, KG representation methods such as TransE [3, 4, 23] and GCN [18] are utilized to encode structural information and embed KGs into low-dimensional spaces individually. Subsequently, for the alignment phase, the embedding spaces are evaluated and compared to derive alignment results under the supervision of seed entity pairs. Certain techniques [7, 14, 15] employ a method of combining training data to create a unified embedding space. This allows for the direct projection of entities from various KGs into the same space. Equivalence across KGs can then be identified by measuring the distance between entities in the unified embedding space during alignment. In order to enhance supervision signals by utilizing the outcomes of the alignment stage, post-alignment iterative techniques are utilized as described in [15, 23]. This approach involves updating structural embeddings and performing alignment recursively until a stopping condition is met. These techniques can be roughly summarized into a framework, depicted by Fig. 6.2.

Recent Advancement on EA Recent endeavors have been directed toward addressing structural heterogeneity by developing sophisticated structural learning

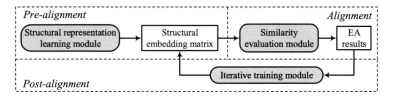

Fig. 6.2 Conventional framework of EA

models such as topic graph matching [21] and multichannel graph neural network [2]. These approaches are intended to overcome the challenges associated with structural heterogeneity. A recent work enhances structural embedding through adversarial training that takes into account degree difference [10]. However, this approach may not be effective when aligning entities in knowledge graphs that are both in a low-frequency range. Furthermore, in this study, degree information is used to improve the learning of structural embeddings, whereas in our approach, degree information is used to combine two different alignment signals: structural and name information.

While iterative strategies can be effective in improving entity alignment (EA), previous research has shown that they can also have drawbacks. For example, they can be biased toward one knowledge graph (KG) and time-consuming [15], or they may introduce many false-positive instances [23], which is not ideal for real-life applications. In order to balance precision and computational efficiency, we propose a novel iterative training approach that incorporates a KG completion module. This module updates the structure of the KG in each round based on confident anchoring entity pairs. Our strategy is lightweight and limits the inclusion of incorrect pairs, reducing the likelihood of introducing false positives.

It is apparent that the majority of the aforementioned embeddings rely on structural information for learning, which can be inadequate for long-tail entities in some cases. To address this issue, some researchers have suggested incorporating *attributes* into embeddings in order to potentially compensate for the shortcomings of relying solely on structural information [14, 17, 18, 22]. However, a significant percentage (between 69 and 99%) of instances in popular KGs are lacking at least one attribute that other entities in the same class possess [6]. The use of *entity descriptions* [3] has been proposed as a way to provide additional information that is often missing in many KGs. While these efforts can improve overall performance, they may not effectively align entities in the long tail. Previous approaches have explored using entity names either as initial features for learning *structural representation* [19–21] or in combination with other information for representation learning [22]. In contrast, our proposed approach consolidates features from separate similarity matrices learned from structure and name information, with different strategies evaluated in Sect. 6.5.2.

6.3   Impact of Long-Tail Phenomenon

Task Definition   Given a source KG $G_1 = (E_1, R_1, T_1)$ and a target KG $G_2 = (E_2, R_2, T_2)$, where E_1 (resp. E_2) represents source (resp. target) entities, R denotes relations and $T \subseteq E \times R \times E$ represents triples. Denote the seed entity pairs as $S = \{(e_1^i, e_2^i)|e_1^i = e_2^i, e_1^i \in E_1, e_2^i \in E_2\}, i \in [1, |S|]$, where $|\cdot|$ denote the cardinality of a set. EA task is to find new EA pairs based on S and return the eventual results $S' = \{(e_1^i, e_2^i)|e_1^i = e_2^i, e_1^i \in E_1, e_2^i \in E_2\}, i \in [1, \min\{|E_1|, |E_2|\}]$, where $=$ expresses that two entities are the same physical one.

A recently published study [7] identified that previous entity alignment datasets had knowledge graphs that were too densely connected and had degree distributions that differed significantly from real-life knowledge graphs. To address this issue, they created a new entity alignment benchmark that better reflects real-life distributions. The benchmark includes both cross-lingual datasets such as SRPRS$_{\text{EN-FR}}$, SRPRS$_{\text{EN-DE}}$, and mono-lingual datasets such as SRPRS$_{\text{DBP-WD}}$ and SRPRS$_{\text{DBP-YG}}$. The degree of an entity is defined as the number of relational triples it participates in. The study reports the degree distributions of entities in the test sets in Table 6.1. The researchers also evaluated the performance of RSNs, which was found to be the best solution in [7]. The evaluation included measuring the number of correctly aligned entities in different degrees.

The results presented in Table 6.1 indicate that in the SRPRS$_{\text{EN-FR}}$ and SRPRS$_{\text{DBP-YG}}$ datasets, over 50% of the entities' degrees are less than three, and in the SRPRS$_{\text{EN-DE}}$ and SRPRS$_{\text{DBP-WD}}$ datasets, almost half of the entities' degree are only 1 or 2. This confirms that the majority of entities in the knowledge graph have very few connections to others and are considered long-tail entities. The results also demonstrate that the accuracy of long-tail entities is much lower than that of higher-degree entities, even though RSNs is the leading method in the benchmark. This suggests that current methods are not effective in handling long-tail entities, which limits overall performance. Therefore, it is crucial to re-evaluate the entity alignment pipeline, with a particular focus on addressing the challenges posed by long-tail entities.

6.4   Methodology

To provide an overview, we have summarized the main components of the DAT (degree-aware entity alignment in tail) framework in Fig. 6.3, highlighting the new designs in purple blue. In pre-alignment, *structural representation learning module* and *name representation learning module* are put forward to learn useful features of entities, i.e., name representation and structural representation; in alignment, these features are forwarded to *degree-aware fusion module* for effective fusion and alignment under the guide of degree information. In post-alignment, *KG completion*

Table 6.1 Degree distribution of entities in test set (the first KG in each KG pair) and results of RSNs

Degree	SRPRS$_{EN-FR}$			SRPRS$_{EN-DE}$			SRPRS$_{DBP-WD}$			SRPRS$_{DBP-YG}$		
	#Total	#Correct	Accuracy	#Total	#Correct	Accuracy	#Total	#Correct	Accuracy	#Total	#Correct	Accuracy
1	2,660	380	14.29%	1,978	453	22.90%	1,341	276	20.58%	3,327	538	16.17%
2	2,540	699	27.52%	2,504	1,005	40.14%	2,979	800	26.85%	2,187	688	31.46%
3	1,130	408	36.11%	1,514	820	54.16%	1,789	600	33.54%	1,143	563	49.26%
>=4	3,120	1,803	57.79%	3,454	2,416	69.95%	3,341	2,093	62.65%	2,793	2,008	71.89%
All	9,450	3,290	34.81%	9,450	4,694	49.67%	9,450	3,769	39.88%	9,450	3,797	40.18%

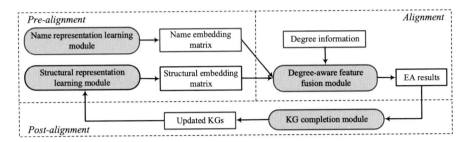

Fig. 6.3 The framework of DAT

module aims to complete KGs with confident EA pairs in the results, and the augmented KGs are then again utilized in the next round iteratively.

Since *structural representation learning module* has been extensively studied, we adopt the state-of-the-art model RSNs [7] for this purpose. Given a structural embedding matrix $\mathbf{Z} \in \mathbb{R}^{n \times d_s}$, two entities $e_1 \in G_1$ and $e_2 \in G_2$, their structural similarity $Sim_s(e_1, e_2)$ is the cosine similarity between $\mathbf{Z}(e_1)$ and $\mathbf{Z}(e_2)$, where n denotes the number of all entities in two KGs, d_s is the dimension of structural embeddings, and $\mathbf{Z}(e)$ denotes the embedding vector for entity e (i.e., $\mathbf{Z}(e) = \mathbf{Z}\mathbf{e}$, where $\mathbf{e}$ is the one-hot encoding of entity e). From the perspective of structure, the target entity with the highest similarity to a source entity is returned as its alignment result.

6.4.1   Name Representation Learning

Remembering that using structural information to align long-tail entities has limited effectiveness, we are taking a different approach from previous attempts that focus on utilizing structures. Instead, we are searching for a signal that is generally accessible to long-tail entities and can provide benefits for alignment.

In order to achieve this goal, we suggest including the textual names of entities, which has largely been ignored by current embedding-based EA methods. This approach is particularly attractive for several reasons, including: (1) the name of an entity is typically sufficient to identify it, and when given two entities, comparing their names is often the most straightforward way to determine if they are equivalent and (2) the majority of real-life entities have a name, and the proportion of entities with names is much greater than the proportion with other textual information, such as descriptions and attributes. This is particularly relevant for long-tail entities, which tend to lack such additional information.

Despite that there are many classic approaches for measuring the *string similarity* between entity names, we go for *semantic similarity* since it can still work when the vocabularies of KGs differ, especially for the *cross-lingual* scenario. Specifically, we choose a general form of power mean embeddings [11], which encompasses many well-known means such as the arithmetic mean, the geometric mean, and the harmonic mean. Given a sequence of word embeddings, $\mathbf{w}_1, \ldots, \mathbf{w}_l \in \mathbb{R}^d$, the power mean operation is formalized as:

$$\left(\frac{w_{1i}^p + \cdots + w_{li}^p}{l} \right)^{1/p}, \quad \forall i = 1, \ldots, d, \quad p \in \mathbb{R} \cup \pm\infty, \tag{6.1}$$

where l is the number of words and d denotes the dimension of embeddings. It can be seen that setting p to 1 results in the arithmetic mean, to 0 the geometric mean, to -1 the harmonic mean, to $+\infty$ the maximum operation, and to $-\infty$ the minimum operation [12].

Given a word embedding space $\mathbb{E}^i$, the embeddings of the words in the name of entity s can be represented as $\mathbf{W}^i = [\mathbf{w}_1^i, \ldots, \mathbf{w}_l^i] \in \mathbb{R}^{l \times d^i}$. Correspondingly, $H_p(\mathbf{W}^i) \in \mathbb{R}^{d^i}$ denotes the power mean embedding vector after feeding $\mathbf{w}_1^i, \ldots, \mathbf{w}_l^i$ to Eq. (6.1). To obtain summary statistics of entity s, we compute K power means of s and concatenate them to get the entity name representation $\mathbf{s}^i \in \mathbb{R}^{d^i \cdot K}$, i.e.,

$$\mathbf{s}^i = H_{p_1}(\mathbf{W}^i) \oplus \cdots \oplus H_{p_K}(\mathbf{W}^i), \tag{6.2}$$

where $\oplus$ represents concatenation along rows and $p_1, \ldots, p_K$ are K different power mean values [12].

To get further representational power from different word embeddings, we generate the final entity name representation $\mathbf{n}_s$ by concatenating $\mathbf{s}^i$ obtained from different embedding spaces $\mathbb{E}^i$:

$$\mathbf{n}_s = \bigoplus_i \mathbf{s}^i. \tag{6.3}$$

Note that the dimensionality of this representation is $d_n = \sum_i d^i \cdot K$. The name embeddings of all entities can be denoted in matrix form as $\mathbf{N} \in \mathbb{R}^{n \times d_n}$.

The representation space will group together entity names that are semantically related, similar to how word embeddings work. When considering the textual names of two entities, denoted as e_1 in group G_1 and e_2 in group G_2, their similarity $Sim_t(e_1, e_2)$ is calculated as the cosine similarity between the vector representation of e_1 and the vector representation of e_2, denoted as $\mathbf{N}(e_1)$ and $\mathbf{N}(e_2)$, respectively. The alignment result for a source entity is the target entity with the highest similarity score.

Discussion The combined power mean word embedding, as presented in the article by Rücklé et al. [12], provides a superior alternative to averaged word embedding when it comes to representing entity names. This is because it is better equipped to capture and synthesize the relevant information conveyed by an entity name.[1] Averaging word embeddings results in a significant loss of information because it fails to account for the semantic variation that can exist within different names. On the other hand, using concatenated power means produces a more accurate summary by reducing ambiguity and uncertainty in the representation of an entity name. This is supported by the empirical evidence presented in Sect. 6.5.3.

It should be noted that in the context of cross-lingual entity alignment, we rely on pre-trained multilingual word embeddings, as described in [5]. These embeddings have already aligned words from different languages into a shared semantic space. As a result, entity names from multiple languages can exist within the same semantic space, obviating the need to design a separate mapping function for aligning multilingual embeddings.

The method described above can be extended to accommodate other textual information, such as attributes, without sacrificing its generality. One simple approach is to concatenate the attributes and entity name to form a "sentence" that provides a more comprehensive description of the entity. This combined sentence can then be encoded using concatenated power mean word embeddings. However, the integration of additional information and more complex adaptations is not within the scope of this chapter.

6.4.2 Degree-Aware Co-attention Feature Fusion

Entity identities can be characterized by various types of features from different perspectives. Therefore, it is important to have a feature fusion module that effectively combines these different signals. Some researchers have proposed to integrate different embeddings into a unified representation space [22], but this approach necessitates additional training to align irrelevant features. A more desirable strategy involves first computing the similarity matrix within each feature-specific space and then combining the similarity scores for each feature-specific space [9, 18]. However, the contributions of different features vary for entities with different degrees. For long-tail entities that lack structural information, entity name representation should be given more weight, whereas for popular entities, the structural representation is relatively more informative than the entity name information. To address this dynamic shift, we draw inspiration from the bi-attention mechanism proposed in [13] and design a degree-aware co-attention network, depicted in Fig. 6.4.

Formally, we are given the structural embedding matrix $\mathbf{Z}$ and the name embedding matrix $\mathbf{N}$. For each entity pair (e_1, e_2), where $e_1 \in G_1$ and $e_2 \in G_2$, we calculate a similarity score between e_1 and e_2. This similarity score is then

---

[1] For possible out-of-vocabulary (OOV) words, we skip them and use the embeddings of the rest to produce entity name embeddings.

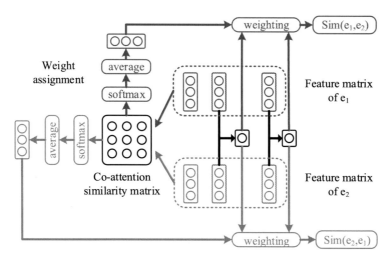

Fig. 6.4 Degree-aware co-attention feature fusion

used to determine the alignment result. To compute the overall similarity between entity pairs, we first calculate the feature-specific similarity scores, $Sim_s(e_1, e_2)$ and $Sim_t(e_1, e_2)$, between e_1 and e_2, as explained in the previous subsections. Our degree-aware co-attention network is designed to determine the weights for $Sim_s(e_1, e_2)$ and $Sim_t(e_1, e_2)$ by incorporating degree information. This network consists of three stages: feature matrix construction, co-attention similarity matrix calculation, and weight assignment.

Feature Matrix Construction Apart from entity name and structural information, we also include entity degree information to construct a feature matrix for each entity. To be precise, we represent entity degrees as one-hot vectors of all possible degree values and pass them through a fully connected layer to obtain a continuous degree vector. As an example, the degree vector of e_1 can be represented as $\mathbf{g}_{e_1} = \mathbf{M} \cdot \mathbf{h}_{e_1} \in \mathbb{R}^{d_g}$, where $\mathbf{h}_{e_1}$ is the one-hot representation of its degree, $\mathbf{M}$ is the weight matrix in the fully-connected layer, and d_g denotes the dimension of the degree vector. This continuous degree vector, along with structural and entity name representations, is stacked to form an entity's feature matrix. For entity e_1:

$$\mathbf{F}_{e_1} = [\mathbf{N}(e_1); \mathbf{Z}(e_1); \mathbf{g}_{e_1}] \in \mathbb{R}^{3 \times d_m}, \tag{6.4}$$

where ; denotes the concatenation along *columns*, $d_m = \max\{d_n, d_s, d_g\}$, and we pad the missing values with 0s.

Co-attention Similarity Matrix Calculation To model the interaction between $\mathbf{F}_{e_1}$ and $\mathbf{F}_{e_2}$, as well as highlight important features, we build a co-attention matrix

$\mathbf{S} \in \mathbb{R}^{3 \times 3}$, where the similarity between the i-th feature of e_1 and the j-th feature of e_2 is computed by:

$$\mathbf{S}_{ij} = \alpha(\mathbf{F}_{e_1}^{i:}, \mathbf{F}_{e_2}^{j:}) \in \mathbb{R}, \qquad (6.5)$$

where $\mathbf{F}_{e_1}^{i:}$ is the i-th row vector and $\mathbf{F}_{e_2}^{j:}$ is the j-th row vector, $i = 1, 2, 3; j = 1, 2, 3$. $\alpha(\mathbf{u}, \mathbf{v}) = \mathbf{w}^{\top}(\mathbf{u} \oplus \mathbf{v} \oplus (\mathbf{u} \circ \mathbf{v}))$ is a trainable scalar function that encodes the similarity, where $\mathbf{w} \in \mathbb{R}^{3d_m}$ is a trainable weight vector and $\circ$ is the element-wise multiplication. Note that the implicit multiplication is a matrix multiplication.

Weight Assignment The co-attention similarity matrix, denoted by $\mathbf{S}$, is used to generate attention vectors, which are $\mathbf{att_1}$ and $\mathbf{att_2}$, in both directions. The attention vector $\mathbf{att_1}$ indicates the feature vectors in e_1 that are most important or relevant to the feature vectors in e_2. Similarly, $\mathbf{att_2}$ indicates the feature vectors in e_2 that are most important or relevant to the feature vectors in e_1. To achieve this, we pass the co-attention similarity matrix $\mathbf{S}$ through a softmax layer. Next, the resulting matrix from the softmax layer is compressed using an average layer to create the attention vectors. It is worth noting that when performing column-wise operations in the softmax layer and row-wise operations in the average layer, we get $\mathbf{att_1}$. Conversely, when conducting row-wise operations in the softmax layer and column-wise operations in the average layer, we obtain $\mathbf{att_2}$.

Eventually, we multiply the feature-specific similarity scores with the attention values to obtain the final similarity score:

$$Sim(e_1, e_2) = Sim_s(e_1, e_2) \cdot \mathbf{att_1}^s + Sim_t(e_1, e_2) \cdot \mathbf{att_1}^t, \qquad (6.6)$$

where $\mathbf{att_1}^s$ and $\mathbf{att_1}^t$ are the corresponding weight values for structural and name similarity scores, respectively. Note that $Sim(e_1, e_2) \neq Sim(e_2, e_1)$ as they may have different attention weight vectors.

The model that combines co-attention and feature fusion has a relatively simple structure with only two parameters, $\mathbf{M}$ and $\mathbf{w}$. Furthermore, it is straightforward to modify this model to include additional features.

Training The training objective is to maximize the similarity scores of the training entity pairs, which can be converted to minimizing the following loss function:

$$L = \sum_{(e_1, e_2) \in S} [-Sim(e_1, e_2) + \gamma]_+ + [-Sim(e_2, e_1) + \gamma]_+, \qquad (6.7)$$

where $[x]_+ = max\{0, x\}$ and and γ is a constant number.

Discussion Alternative methods of implementing degree-aware weighting are possible, such as applying `sigmoid`$(\mathbf{W} \cdot [\mathbf{N}(e), \mathbf{Z}(e), \mathbf{g}_e])$ where $\mathbf{W}$ represents the parameter. In this study, we utilize a co-attention mechanism to combine various signal channels with degree-aware weights, which highlights the benefits of incorporating degrees for effective EA in the tail. However, a more comprehensive comparison with other implementations is a subject for future research.

6.4.3 Iterative KG Completion

The concept of iterative self-training has been shown to be effective and warrants further investigation, as demonstrated in previous studies [15, 23]. However, current research has failed to consider the potential for enriching structural information during the iterative process. Our findings suggest that, while long-tail entities in the source KG may lack structural information, this information can be found in the target KG in a complementary manner. By mining confident EA results and using them as pseudo matching pairs to anchor subgraphs, we can replenish the original KG with facts from its counterpart, thereby mitigating the KGs' structural sparsity. This can significantly improve KG coverage and reduce the number of long-tail entities. As the structural learning model generates increasingly better structural embeddings from the amplified KGs, the accuracy of EA results in subsequent rounds also improves naturally in an iterative fashion.

To start, we will describe how we incorporate EA pairs that have a high level of confidence. Our focus is on preventing the inclusion of any incorrect pairs that could potentially harm the model. To achieve this, we have developed a unique approach for choosing EA pairs. For every given entity $e_1 \in E_1 - S_1$ (in G_1 but not in the training set), suppose its most similar entity in G_2 is e_2, its second most similar entity is e_2' and the difference between the similarity scores is $\Delta_1 \triangleq Sim(e_1, e_2) - Sim(e_1, e_2')$, if for e_2, its most similar entity in G_1 is exactly e_1, its second most similar entity is e_1', the difference between the similarity scores is $\Delta_2 \triangleq Sim(e_2, e_1) - Sim(e_2, e_1')$, and Δ_1, Δ_2 are both above a given threshold θ, (e_1, e_2) would be considered as a correct pair. This is a relatively strong constraint, as it requires that (1) the similarity between the two entities is the highest from both sides, respectively, and (2) there is a margin between the top two candidates.

Once we have integrated the EA results with high confidence to the initial set of entity pairs, we proceed to use these entities (S_a) to connect two KGs and supplement them with new facts from each other. For example, if a triple $t_1 \in T_1$ has both its head and tail entities matching entries in S_a, we replace the entities in t_1 with the corresponding entities in E_2 and add the new triple to T_2. While this may seem like a simple and straightforward approach, it effectively increases the overall coverage of the KGs. Finally, we leverage the augmented KGs to improve the quality of the structural representations, which in turn contributes to enhancing the EA performance. This iterative completion process is repeated for ζ rounds.

Discussion Certain EA methods also use bootstrapping or iterative training techniques, but their primary goal is to expand the training signals for updating the embeddings, without modifying the underlying structure of the KGs. In comparison to other approaches of selecting EA pairs which can be slow and may generate inaccurate results [15, 23], we improve this process by prioritizing two entities if they give each other priority. This is empirically validated in Sect. 6.5.5.

6.5 Experiments

This section reports the experiments with in-depth analysis.[2]

6.5.1 Experimental Setting

Dataset We use SRPRS [7] due to the KG pairs having a distribution similar to the real world. It was created with inter-language links and references in DBpedia, and each entity has an equivalent counterpart in the other KG. The relevant details are listed in Table 6.2, and 30% of entity pairs are utilized for training.

Parameter Settings For the *structural representation learning module*, we follow the settings in [7], except for assigning d_s to 300. Regarding *name representation learning module*, we set $\mathbf{p} = [p_1, \ldots, p_K]$ to [1, min, max]. For mono-lingual datasets, we merely use the fastText embeddings [1] as the word embedding (i.e., only one embedding space in Eq. (6.3)). For cross-lingual datasets, the multilingual word embeddings are obtained from MUSE.[3] Two word embedding spaces (from two languages) are used in Eq. (6.3). As for *degree-aware fusion module*, we set d_g to 300, γ to 0.8, and batch size to 32. Stochastic gradient descent is harnessed to minimize the loss function, with learning rate set to 0.1, and we use early stopping to prevent over-fitting. In *KG completion module*, θ is set to 0.05 and ζ is set to 3.

Evaluation Metric We use Hits@k ($k = 1, 10$) and the mean reciprocal rank (MRR) as evaluation metrics. For each source entity, entities in the other KG are

Table 6.2 Statistics of SRPRS

Dataset	KGs	#Triples	#Entities
SRPRS$_{EN-FR}$	DBpedia (English)	36,508	15,000
	DBpedia (French)	33,532	15,000
SRPRS$_{EN-DE}$	DBpedia (English)	38,281	15,000
	DBpedia (German)	37,069	15,000
SRPRS$_{DBP-WD}$	DBpedia	38,421	15,000
	Wikidata	40,159	15,000
SRPRS$_{DBP-YG}$	DBpedia	33,571	15,000
	YAGO3	34,660	15,000

---

[2] The source code is available at https://github.com/DexterZeng/DAT.

[3] https://github.com/facebookresearch/MUSE.

ranked according to their similarity scores Sim with the source entity in descending order. Hits@k measures the proportion of correctly aligned entities among the top-k similar entities to the source entity. In particular, Hit@1 indicates the accuracy of the alignment results. MRR, on the other hand, is the average of the reciprocal ranks of the ground-truth results. A higher Hits@k and MRR indicate better performance. Unless stated otherwise, the results of Hits@k are represented as percentages. The best results are displayed in **bold** in the tables.

Competitors Overall 13 state-of-the-art methods are involved in comparison. The group that solely utilizes structural feature includes (1) MTransE [4], which proposes to utilize TransE for EA; (2) IPTransE [23], which uses an iterative training process to improve the alignment results; (3) BootEA [15], which devises an alignment-oriented KG embedding framework and a bootstrapping strategy; (4) RSNs [7], which integrates recurrent neural networks with residual learning; (5) MuGNN [2], which puts forward a multichannel graph neural network to learn alignment-oriented KG embeddings; (6) KECG [8], which proposes to jointly learn knowledge embeddings that encode inner-graph relationships, and a cross-graph model that enhances entity embeddings with their neighbors' information; and (7) TransEdge [16], which presents a novel edge-centric embedding model that contextualizes relation representations in terms of specific head-tail entity pairs.

Various methods have been proposed to incorporate other types of information in EA. JAPE [14] utilizes attributes of entities to refine structural information. GCN [18] generates entity embeddings and attribute embeddings to align entities in different KGs. GM-Align [21] builds a local subgraph of an entity to represent it and utilizes entity name information to initialize the framework. MultiKE [22] offers a novel framework that unifies the views of entity names, relations, and attributes at *representation-level for mono-lingual* EA. RDGCN [19] proposes a relation-aware dual-graph convolutional network to incorporate relation information via attentive interactions between KG and its dual relation counterpart. HGCN [20] is a learning framework that jointly learns entity and relation representations for EA.

6.5.2 Results

Table 6.3 presents the results. The first group of approaches only use structural information for alignment. BootEA and KECG outperform MTransE and IPTransE because of their alignment-oriented KG embedding framework and attention-based graph embedding model, respectively. RSNs further improves the results by taking into account long-term relational dependencies between entities, which can capture more structural signals for alignment. TransEdge achieves the best performance due to its edge-centric KG embedding and bootstrapping strategy. MuGNN fails to produce effective results as there are no aligned relations on SRPRS, which prevents the rule transferring from taking place and limits the number of detected rules. It is noteworthy that Hits@1 values on most datasets are below 50%, demonstrating

Table 6.3 Overall results of entity alignment

Methods	SRPRS_EN-FR			SRPRS_EN-DE			SRPRS_DBP-WD			SRPRS_DBP-YG		
	Hits@1	Hits@10	MRR	Hits@1	Hits@10	MRR	Hits@1	Hits@10	MRR	Hits@1	Hits@10	MRR
MTransE	25.1	55.1	0.35	31.2	58.6	0.40	22.3	50.1	0.32	24.6	54.0	0.34
IPTransE	25.5	55.7	0.36	31.3	59.2	0.41	23.1	51.7	0.33	22.7	50.0	0.32
BootEA	31.3	62.9	0.42	44.2	70.1	0.53	32.3	63.1	0.42	31.3	62.5	0.42
RSNs	34.8	63.7	0.45	49.7	73.3	0.58	39.9	66.8	0.49	40.2	68.9	0.50
MuGNN	13.1	34.2	0.20	24.5	43.1	0.31	15.1	36.6	0.22	17.5	38.1	0.24
KECG	29.8	61.6	0.40	44.4	70.7	0.54	32.3	64.6	0.43	35.0	65.1	0.45
TransEdge	40.0	67.5	0.49	55.6	75.3	0.63	46.1	73.8	0.56	44.3	69.9	0.53
GCN	15.5	34.5	0.22	25.3	46.4	0.33	17.7	37.8	0.25	19.3	41.5	0.27
JAPE	25.6	56.2	0.36	32.0	59.9	0.41	21.9	50.1	0.31	23.3	52.7	0.33
RDGCN	67.5	76.9	0.71	78.3	88.4	0.82	83.4	90.7	0.86	85.8	93.8	0.89
HGCN	67.0	77.0	0.71	76.3	86.3	0.80	82.3	88.7	0.85	82.2	88.8	0.85
GM-Align[a]	62.7	–	–	67.7	–	–	81.5	–	–	82.8	–	–
DAT	**75.8**	**89.9**	**0.81**	**87.6**	**95.5**	**0.90**	**92.6**	**97.7**	**0.94**	**94.0**	**98.5**	**0.96**

[a] When running GM-Align, it is noted that entities without valid name embeddings are excluded from evaluation, and hence we consider that GM-Align fails to align these entities without specifying rankings, which leads to the lack of Hits@10 and MRR values

the inadequacy of solely relying on KG structure, especially when long-tail entities make up the majority.

Regarding the second group, both GCN and JAPE exploit attribute information to complement structural signals. However, they fail to outperform the leading method in the first group, which can be attributed to the limited effect of attributive information. The other four methods make use of the publicly available entity name data. The substantial improvement in results compared to those of the first group confirms the value of this feature. Our framework, DAT, demonstrates its superiority over GM-Align, RDGCN, and HGCN with a 10% improvement in Hits@1 over all datasets, validating the effectiveness of exploiting entity name information. The fundamental explanation for this is that the fusion of features on the representation level by GM-Align, RDGCN, and HGCN may lead to information loss since the resulting merged feature representation may not retain the distinguishing features of the original ones. On the other hand, DAT adopts a co-attention network to compute feature weights and fuse features at the output level, which is based on feature-specific similarity scores.

Evaluation by Degree We present the outcomes of DAT in terms of degree to illustrate its ability to align long-tail entities, as shown in Table 6.4. It is worth noting that the degree pertains to the original degree distribution since the entity degree may be changed by the completion process.

Table 6.4 indicates that for entities with a degree of 1, the Hits@1 scores of DAT are two or three times higher than those of RSNs, confirming the capability of DAT in handling the long-tail problem. While there is also an improvement in the performance of DAT for popular entities, the gap between DAT and RSNs is much smaller than that observed in the case of long-tail entities. Furthermore, DAT outperforms RDGCN in all degree categories across four datasets, despite both using entity name information as an external signal for EA.

Comparison with MultiKE on Dense Datasets The reason for not providing the results of MultiKE on SRPRS is because it can only handle datasets in a single language and requires prior knowledge of the relations' semantics. However, in order to better understand DAT, we present the experimental results of DAT on the *dense* datasets that were previously evaluated with MultiKE. Specifically, the dense datasets, $DWY100K_{DBP-WD}$ and $DWY100K_{DBP-YG}$, are similar to $SRPRS_{DBP-WD}$ and $SRPRS_{DBP-YG}$, but have a larger scale (100K entities on each side) and higher density [15].

When evaluated on dense datasets, DAT produces superior results with Hits values exceeding 90% and MRR surpassing 0.95, as presented in Table 6.5. This indicates that DAT effectively utilizes name information, which can be credited to the degree-aware feature fusion module and the approach of first computing scores within each view rather than learning a merged representation that may result in the loss of information.

Table 6.4 Hits@1 results by degrees

Degree	SRPRS$_{EN-FR}$			SRPRS$_{EN-DE}$			SRPRS$_{DBP-WD}$			SRPRS$_{DBP-YG}$		
	RSNs	RDGCN	DAT	RSNs	RDGCN	DAT	RSNs	RDGCN	DAT	RSNs	RDGCN	DAT
1	14.3	56.5	**57.4**	22.9	75.8	**83.1**	20.6	80.3	**86.7**	16.2	84.2	**89.5**
2	27.5	64.4	**72.4**	40.1	78.6	**84.3**	26.9	84.0	**90.8**	31.5	80.9	**92.6**
3	36.1	77.3	**82.9**	54.2	77.4	**88.4**	33.5	75.2	**88.2**	49.3	89.2	**97.0**
≥4	57.8	75.8	**91.6**	69.9	80.0	**92.3**	62.6	88.6	**98.8**	71.9	90.2	**99.4**
All	34.8	67.5	**75.8**	49.7	78.3	**87.6**	39.9	83.4	**92.6**	40.2	85.8	**94.0**

Table 6.5 Experimental results on dense datasets

Methods	DWY100K$_{DBP-WD}$			DWY100K$_{DBP-YG}$		
	Hits@1	Hits@10	MRR	Hits@1	Hits@10	MRR
MultiKE	91.9	96.3	0.94	88.0	95.3	0.91
DAT	**97.4**	**99.6**	**0.98**	**94.3**	**98.6**	**0.96**

Table 6.6 Experimental results of ablation

Methods	SRPRS$_{EN-FR}$		
	Hits@1	Hits@10	MRR
DAT	**75.8**	**89.9**	**0.81**
DAT w/o IKGC	72.1	85.4	0.77
DAT w/o KGC	73.9	88.6	0.79
DAT w/o ATT	73.1	88.5	0.79
DAT w/o CPM	75.3	89.7	0.80

Fig. 6.5 Distribution of entity degree in SRPRS$_{EN-FR}$

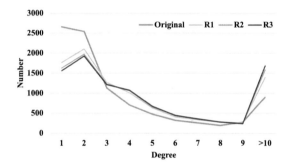

6.5.3 Ablation Study

We report an ablation study on SRPRS$_{EN-FR}$ dataset in Table 6.6.

Iterative KG Completion If we remove the entire module, the performance of EA drops by 3.7% on Hits@1 (comparing DAT with DAT w/o IKGC). However, if we eliminate only the KG completion module while keeping the iterative process (similar to [23]), Hits@1 decreases by 1.9% (DAT vs. DAT w/o KGC). This validates the significance of KG completion. We also present the dynamic change of the degree distribution after each round (original, R1, R2, R3) in Fig. 6.5, which suggests that the embedded KG completion improves KG coverage and reduces the number of long-tail entities.

Degree-Aware Co-attention Feature Fusion In Table 6.6, it can be observed that if the fixed equal weights are used instead of the *degree-aware fusion module*, the Hits@1 decreases by 2.7% (DAT vs. DAT w/o ATT). This result confirms that adjusting the weights of features dynamically based on their degree leads to better integration of features and, as a result, more accurate alignment results. In Fig. 6.6, we present the weight of the structural representation generated by our degree-aware

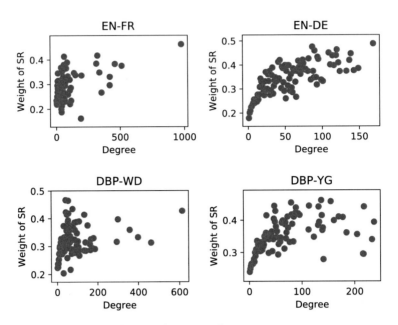

Fig. 6.6 Weight distribution of structural representation

fusion model across different degrees (in the first round). This figure demonstrates that, in general, the importance of structural information increases with the degree of entities, which is in line with our expectations.

Concatenated Power Mean Word Embeddings We compared concatenated power mean word embeddings and averaged word embeddings in terms of aligning entities, denoting as DAT and DAT w/o CPM, respectively. The findings indicate that combining multiple power mean embeddings effectively captures more alignment features.

6.5.4   Error Analysis

We conduct an error analysis on $\mathrm{SRPRS}_{\mathrm{EN-FR}}$ dataset to investigate the contribution of each module and cases where DAT falls short. Using only structural information leads to a high error rate of 65.5% on Hits@1. The dataset contains 67.0% long-tail(i.e., with degree ≤ 3) entities, with a majority (65.1%) being misaligned. However, incorporating entity name information and dynamically fusing it with structural information significantly reduces the overall Hits@1 error rate to 27.9%, with a corresponding reduction in long-tail entity error rate to 33.2%. Furthermore, we employ iterative KG completion to replenish structure and propagate signals, which further decrease the overall Hits@1 error rate to 24.2%. This approach also reduces the percentage of long-tail entities to 49.7%, with only 8.3% being misaligned. Overall, our results indicate that long-tail entities initially account for the most errors, but employing the proposed techniques reduces not only the error rate but also the contribution of long-tail entities to the overall error.

For cases that DAT cannot solve, we provide an analysis that focuses on the information related to entity names. Out of the incorrect cases (24.2% in $SRPRS_{EN-FR}$), 41% don't have an appropriate entity name embedding because all the words in the name are out-of-vocabularies (OOVs), and 31% have partial OOVs. Additionally, 15% could have been correct by solely utilizing the name information, but they were misled by structural signals, while 13% fail to align because of either the inadequacy of the entity name representation method or the fact that the entities with the same name refer to different physical objects.

6.5.5 Further Experiment

We substantiate the efficacy of our iterative training approach by performing the following experiments.

Our iterative approach differs from current methods not only in the embedded KG completion procedure but also in the choice of confident pairs. To showcase its advantage, we remove the KG completion module from DAT and obtain DAT-I to compare its selection methods with those of [15, 23]. In [23], the authors use a threshold-based method (TH) to find pairs. For each *nonaligned* source entity, it identifies the most comparable *nonaligned* target entity, and if the similarity between the two entities exceeds a specified threshold, they are deemed confident pairs. In [15], the authors use a maximum weight graph matching (MWGM) method to find confident entity alignment pairs. For each source·entity, it calculates the alignment likelihood to every target entity, and only those with likelihood above a given threshold are considered in a maximum likelihood matching process under a 1-to-1 mapping constraint, which generates a solution that contains confident EA pairs. We implement the methods within our framework and adjust the parameters based on the original papers. To evaluate the effectiveness of various iterative training techniques, we use the number of chosen confident EA pairs, the accuracy of these pairs, and the duration of each round as primary metrics.

To ensure fairness in the comparison, we present the outcomes of the initial three rounds in Fig. 6.7. The findings indicate that DAT-I outperforms the other two methods regarding the quantity and quality of chosen pairs in a relatively shorter time. As MWGM necessitates solving a global optimization problem, it takes considerably more time. Nonetheless, compared to TH, it performs better in terms of the accuracy of selected pairs.

Fig. 6.7 Comparison results of iterative training strategies. (**a**) Number of pairs selected. (**b**) Accuracy of pairs selected. (**c**) Running time consumption (s)

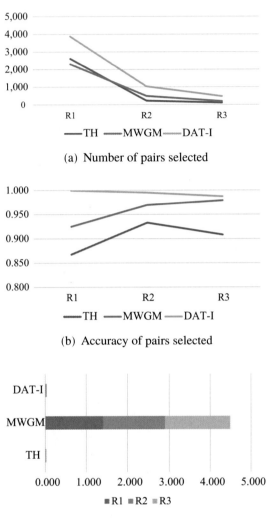

(a) Number of pairs selected

(b) Accuracy of pairs selected

(c) Running time consumption (s)

6.6 Conclusion

In this chapter, we present an improved framework called DAT for entity alignment, which specifically focuses on handling long-tail entities. Recognizing the limitations of relying solely on structural information, we propose to incorporate entity name information in the pre-alignment phase through concatenated power mean embedding. For alignment, we introduce a co-attention feature fusion network that dynamically adjusts the weights of different features guided by degree to consolidate various signals. In the post-alignment phase, we enhance the performance by iteratively completing the KG with confident EA results as anchors, thereby amplifying the structural information. We evaluate DAT on cross-lingual and monolingual EA benchmarks and achieve superior results.

References

1. P. Bojanowski, E. Grave, A. Joulin, and T. Mikolov. Enriching word vectors with subword information. *Transactions of the Association for Computational Linguistics*, 5:135–146, 2017.
2. Y. Cao, Z. Liu, C. Li, Z. Liu, J. Li, and T. Chua. Multi-channel graph neural network for entity alignment. In *Proceedings of the 57th Conference of the Association for Computational Linguistics, ACL 2019, Florence, Italy, July 28- August 2, 2019, Volume 1: Long Papers*, pages 1452–1461, 2019.
3. M. Chen, Y. Tian, K. Chang, S. Skiena, and C. Zaniolo. Co-training embeddings of knowledge graphs and entity descriptions for cross-lingual entity alignment. In *Proceedings of the Twenty-Seventh International Joint Conference on Artificial Intelligence, IJCAI 2018, July 13-29, 2018, Stockholm, Sweden.*, pages 3998–4004, 2018.
4. M. Chen, Y. Tian, M. Yang, and C. Zaniolo. Multilingual knowledge graph embeddings for cross-lingual knowledge alignment. In *Proceedings of the Twenty-Sixth International Joint Conference on Artificial Intelligence, IJCAI 2017, Melbourne, Australia, August 19-25, 2017*, pages 1511–1517, 2017.
5. A. Conneau, G. Lample, M. Ranzato, L. Denoyer, and H. Jégou. Word translation without parallel data. *CoRR*, abs/1710.04087, 2017.
6. L. Galárraga, S. Razniewski, A. Amarilli, and F. M. Suchanek. Predicting completeness in knowledge bases. In *Proceedings of the Tenth ACM International Conference on Web Search and Data Mining, WSDM 2017, Cambridge, United Kingdom, February 6-10, 2017*, pages 375–383, 2017.
7. L. Guo, Z. Sun, and W. Hu. Learning to exploit long-term relational dependencies in knowledge graphs. In *Proceedings of the 36th International Conference on Machine Learning, ICML 2019, 9-15 June 2019, Long Beach, California, USA*, pages 2505–2514, 2019.
8. C. Li, Y. Cao, L. Hou, J. Shi, J. Li, and T.-S. Chua. Semi-supervised entity alignment via joint knowledge embedding model and cross-graph model. In *Proceedings of the 2019 Conference on Empirical Methods in Natural Language Processing and the 9th International Joint Conference on Natural Language Processing (EMNLP-IJCNLP)*, pages 2723–2732, 2019.
9. N. Pang, W. Zeng, J. Tang, Z. Tan, and X. Zhao. Iterative entity alignment with improved neural attribute embedding. In *Proceedings of the Workshop on Deep Learning for Knowledge Graphs (DL4KG2019) Co-located with the 16th Extended Semantic Web Conference 2019 (ESWC 2019), Portoroz, Slovenia, June 2, 2019.*, pages 41–46, 2019.
10. S. Pei, L. Yu, R. Hoehndorf, and X. Zhang. Semi-supervised entity alignment via knowledge graph embedding with awareness of degree difference. In *The World Wide Web Conference, WWW 2019, San Francisco, CA, USA, May 13-27, 2019*, pages 3130–3136, 2019.
11. G. Polya, G. H. Hardy, and Littlewood. *Inequalities*. University Press, 1952.
12. A. Rücklé, S. Eger, M. Peyrard, and I. Gurevych. Concatenated p-mean word embeddings as universal cross-lingual sentence representations. *CoRR*, abs/1803.01400, 2018.
13. M. J. Seo, A. Kembhavi, A. Farhadi, and H. Hajishirzi. Bidirectional attention flow for machine comprehension. In *5th International Conference on Learning Representations, ICLR 2017, Toulon, France, April 24-26, 2017, Conference Track Proceedings*, 2017.
14. Z. Sun, W. Hu, and C. Li. Cross-lingual entity alignment via joint attribute-preserving embedding. In *The Semantic Web - ISWC 2017 - 16th International Semantic Web Conference, Vienna, Austria, October 21-25, 2017, Proceedings, Part I*, pages 628–644, 2017.

15. Z. Sun, W. Hu, Q. Zhang, and Y. Qu. Bootstrapping entity alignment with knowledge graph embedding. In *Proceedings of the Twenty-Seventh International Joint Conference on Artificial Intelligence, IJCAI 2018, July 13-29, 2018, Stockholm, Sweden.*, pages 4396–4402, 2018.
16. Z. Sun, J. Huang, W. Hu, M. Chen, L. Guo, and Y. Qu. Transedge: Translating relation-contextualized embeddings for knowledge graphs. In *The Semantic Web - ISWC 2019 - 18th International Semantic Web Conference, Auckland, New Zealand, October 26-30, 2019, Proceedings, Part I*, pages 612–629, 2019.
17. B. D. Trisedya, J. Qi, and R. Zhang. Entity alignment between knowledge graphs using attribute embeddings. In *The Thirty-Third AAAI Conference on Artificial Intelligence, AAAI 2019, The Thirty-First Innovative Applications of Artificial Intelligence Conference, IAAI 2019, The Ninth AAAI Symposium on Educational Advances in Artificial Intelligence, EAAI 2019, Honolulu, Hawaii, USA, January 27 - February 1, 2019.*, pages 297–304, 2019.
18. Z. Wang, Q. Lv, X. Lan, and Y. Zhang. Cross-lingual knowledge graph alignment via graph convolutional networks. In *Proceedings of the 2018 Conference on Empirical Methods in Natural Language Processing, Brussels, Belgium, October 31 - November 4, 2018*, pages 349–357, 2018.
19. Y. Wu, X. Liu, Y. Feng, Z. Wang, R. Yan, and D. Zhao. Relation-aware entity alignment for heterogeneous knowledge graphs. In *Proceedings of the Twenty-Eighth International Joint Conference on Artificial Intelligence, IJCAI 2019, Macao, China, August 10-16, 2019*, pages 5278–5284, 2019.
20. Y. Wu, X. Liu, Y. Feng, Z. Wang, and D. Zhao. Jointly learning entity and relation representations for entity alignment. In *Proceedings of the 2019 Conference on Empirical Methods in Natural Language Processing and the 9th International Joint Conference on Natural Language Processing (EMNLP-IJCNLP)*, pages 240–249, 2019.
21. K. Xu, L. Wang, M. Yu, Y. Feng, Y. Song, Z. Wang, and D. Yu. Cross-lingual knowledge graph alignment via graph matching neural network. In *Proceedings of the 57th Conference of the Association for Computational Linguistics, ACL 2019, Florence, Italy, July 28- August 2, 2019, Volume 1: Long Papers*, pages 3156–3161, 2019.
22. Q. Zhang, Z. Sun, W. Hu, M. Chen, L. Guo, and Y. Qu. Multi-view knowledge graph embedding for entity alignment. In *Proceedings of the Twenty-Eighth International Joint Conference on Artificial Intelligence, IJCAI 2019, Macao, China, August 10-16, 2019*, pages 5429–5435, 2019.
23. H. Zhu, R. Xie, Z. Liu, and M. Sun. Iterative entity alignment via joint knowledge embeddings. In *Proceedings of the Twenty-Sixth International Joint Conference on Artificial Intelligence, IJCAI 2017, Melbourne, Australia, August 19-25, 2017*, pages 4258–4264, 2017.
24. Q. Zhu, X. Zhou, J. Wu, J. Tan, and L. Guo. Neighborhood-aware attentional representation for multilingual knowledge graphs. In *Proceedings of the Twenty-Eighth International Joint Conference on Artificial Intelligence, IJCAI 2019, Macao, China, August 10–16, 2019*, pages 1943–1949, 2019.

Chapter 7
Weakly Supervised Entity Alignment

Abstract The majority of state-of-the-art entity alignment solutions heavily rely on the labeled data, which are difficult to obtain in practice. Therefore, it calls for the study of EA with *scarce supervision*. To resolve this issue, we put forward a *reinforced active* entity alignment framework to select the entities to be manually labeled with the aim of enhancing alignment performance with minimal labeling efforts. Under this framework, we further devise an unsupervised *contrastive* loss to contrast different views of entity representations and augment the limited supervision signals by exploiting the vast unlabeled data. We empirically evaluate our proposal on eight popular KG pairs, and the results demonstrate that our proposed model and its components consistently boost the alignment performance under scarce supervision.

7.1 Introduction

The entity alignment performance heavily relies on the amount of labeled data (i.e., aligned entity pairs). It has been empirically verified that the alignment accuracy drops sharply when decreasing the number of seed entity pairs [23]. This is also illustrated in Fig. 7.1, where we summarize the alignment performance of the most performant EA solutions given varying sizes of training data.[1] Although this problem is prominent, it has been largely neglected by existing literature, since they directly extract the supervision signals from the inter-language links in DBpedia [1] or reference links among DBpedia, YAGO [22], Freebase [3], and Wikidata [28]. In practice though, these prior alignments might not exist among KGs constructed from different sources. In this case, it requires manual annotation to produce such labeled data, which is a nontrivial task since the annotator needs to retrieve the entity

---

[1] Note that in this chapter, we confine the main discussion to EA solutions that only use the KG structure and exclude those using auxiliary information such as entity descriptions, since some KGs might have little or even no auxiliary information [35], and the former can be regarded as a more general case of EA. Nevertheless, we will show that our proposed method can also be applied to the latter in the experiment.

© The Author(s) 2023
X. Zhao et al., *Entity Alignment*, Big Data Management,
https://doi.org/10.1007/978-981-99-4250-3_7

Fig. 7.1 The Hits@1
alignment results of
state-of-the-art methods
(RREA [17], MRAEA [16],
SSP [19], TransEdge [24],
and AliNet [25]) on
DBP15K$_{ZH-EN}$ [23] given
decreasing amount of labeled
data. Label rate denotes the
percentage of labeled data in
the whole dataset

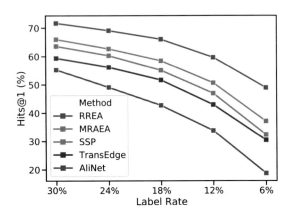

equivalent to a given entity from a vast pool of candidates. As thus, to reduce the manual labeling cost and also the reliance on labeled data, it is of great significance to study EA with scarce supervision.

In this chapter, we propose to approach EA with limited supervision by addressing two key research questions: **(Q1)** Given a fixed labeling budget, how to select the entities for manual annotation so that the labeled data can provide more useful guidance for the alignment? This can also be interpreted as, to reach a certain target of alignment performance, how to optimize the selection of entities for labeling so that we could label as few entities as possible? **(Q2)** Given a limited number of labeled data, how can we leverage the rich unlabeled data to facilitate the alignment?

In response to **Q1**, we exploit active learning (AL) to overcome the labeling bottleneck by asking queries in the form of unlabeled entities to be labeled by an oracle (e.g., a human annotator) [9, 21]. Through designing effective query strategies, the active learner can achieve satisfying performance using as few labeled instances as possible, thereby reducing the cost of obtaining labeled data [21]. In this chapter, we develop several query strategies to characterize the informativeness of entities from different angles and offer a reinforced AL framework to blend these query strategies adaptively with the aim of selecting the most valuable entities to be labeled.

To answer **Q2**, inspired by the recent advance in contrastive learning (CL) [12, 27], we devise an unsupervised contrastive loss to exploit the abundant unlabeled data for augmenting supervision signals. CL aims to generate data representations by learning to encode the similarities or dissimilarities among a set of unlabeled examples. The underlying intuition is that the rich unlabeled data themselves can be used as supervision to help guide the model training [29]. In this chapter, we employ two graph encoders to model different views of the structural information of entities and design a contrastive objective to distinguish the embeddings of the same entity in these two views from the embeddings of other entities. By incorporating the unsupervised contrastive loss into the semi-supervised alignment objective, the scarce supervision signals are amplified, which can further ameliorate the alignment performance.

The reinforced AL and the contrastive representation learning constitute RAC, an EA framework developed specifically to deal with scarce supervision. We empirically evaluate RAC on eight popular KG pairs against active and non-active baseline models. The results demonstrate that RAC achieves superior performance under scarce supervision and can be applied on existing EA models.

Contribution The main contribution of this chapter can be summarized as follows: (1) we put forward RAC, an EA framework that aims to solve the scarce supervision issue, which can be employed on top of existing EA solutions to improve their capability of tackling limited labeled data; (2) we devise a reinforced active learning approach to blend query heuristics adaptively and select valuable entities for labeling, which benefits the subsequent alignment process; and (3) we are among the first attempts to exploit contrastive learning for EA, where the underlying supervision signals in the abundant unlabeled entities are leveraged to facilitate alignment.

7.2 Preliminaries

7.2.1 Problem Formulation

A KG is denoted as $G = \{(s, r, o)\} \subset \mathcal{E} \times \mathcal{R} \times \mathcal{E}$, where $\mathcal{E}$ is the entity set, $\mathcal{R}$ is the relation set, and a triple (s, r, o) represents a subject entity $s \in \mathcal{E}$ and an object entity $o \in \mathcal{E}$ are connected by a relation $r \in \mathcal{R}$. The inputs to entity alignment include two KGs to be aligned (i.e., the source KG G_s and the target KG G_t), a set of labeled entity pairs $S = \{(u^*, v^*)|u^* \in \mathcal{E}_s, v^* \in \mathcal{E}_t, u^* \Leftrightarrow v^*\}$, where $\Leftrightarrow$ represents equivalence, $\mathcal{E}_s$ and $\mathcal{E}_t$ refer to the entity sets in the source and target KGs, respectively. The objective is to detect equivalent entity pairs in the rest of the entities.

The focus of this chapter is to study entity alignment with *scarce supervision*, which is decomposed into two problems, i.e., selecting entities for labeling and entity alignment under such limited supervision signals. The former is defined as, given a pool of unlabeled entities $\mathcal{U}$, a labeling budget B and an oracle to label the entities, selecting B entities from $\mathcal{U}$ for annotation so that the labeled data could provide more useful guidance for the subsequent alignment.

7.2.2 Model Overview

We provide the overview of our proposed model RAC in Fig. 7.2. RAC can be decomposed into multiple iterations. In each iteration, we first conduct **reinforced**

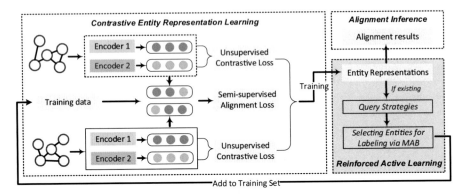

Fig. 7.2 The framework of our proposal RAC

active learning, where we use query strategies to measure the informativeness of entities and exploit the multi-armed bandit (MAB) mechanism to blend these strategies adaptively, so as to produce the entities to be labeled by the oracle. Next, the labeled entity pairs are added to the training set and forwarded to the **contrastive entity representation learning** module. In this module, using the labeled entity pairs as connections, we project individual KGs to a unified embedding space, where the entities from different KGs become comparable and the equivalence between entities can thus be inferred. Specifically, we design a semi-supervised alignment loss function to enforce the embeddings of the entities in each labeled entity pair to be close, such that the supervision signals can be propagated to unlabeled entities and their embeddings are updated to be comparable across KGs. Then we supplement with the unsupervised contrastive loss to contrast the structural entity representations learned by different graph encoders, which can leverage the rich unlabeled information for learning more expressive entity representations. Finally, the learned unified entity representations are used to conduct **alignment inference** to produce the results and also to help improve query strategies.

7.3 Reinforced Active Learning

To cope with EA with scarce supervision, we first address the entity selection (for annotation) problem. Concretely, we adopt active learning (AL) to select the entities to be manually labeled with the aim of maximizing model performance with minimal effort. The AL process normally consists of multiple iterations. Given the labeling budget B, in each iteration, guided by the query strategies, we select $b(b < B)$ entities with the highest *informativeness* for labeling and add the annotated entity pairs into the labeled data for training the EA framework. The iteration continues until the labeling budget is exhausted. Next, we introduce the query strategies in detail and then elaborate the reinforced active entity selection framework.

7.3.1 Query Strategies

We leverage three query strategies, i.e., degree centrality, PageRank centrality, and information density, to characterize the *informativeness* (more specifically, the representativeness) of entities.[2]

Degree and PageRank Centrality Since the entities in KGs are not i.i.d., we consider nodes with higher centrality contain more useful information and are of greater values. Hence, we adopt the commonly used centrality metric, i.e., the degree centrality $\phi_{deg}(e)$, which is defined as the number of edges directly connected with entity e. Besides, we also leverage the PageRank centrality [18] $\phi_{pr}(e_i; \Theta_p)$ to characterize the representativeness of entities:

$$\phi_{pr}(e_i; \Theta_p) = \rho \sum_j A_{ij} \frac{\phi_{pr}(e_j; \Theta_p)}{\sum_k A_{jk}} + \frac{1 - \rho}{n}, \tag{7.1}$$

where A is the adjacency matrix, n is the number of entities in the KG, ρ is the damping parameter, and Θ_p denotes the parameter set.

Information Density In addition to the topological structure, the representativeness of an entity can also be measured from the embedding level. Concretely, we apply K-means on the embeddings of unlabeled entities. We consider the entities placed at or close to the centers of clusters are of greater values. Thus, we calculate the Euclidean distance $d(e, c_e)$ between each entity e and the center entity c_e of the cluster it belongs to and define the *information density* of entity e as $\phi_i(e; \Theta_i) = \frac{1}{1+d(e, c_e)}$, where Θ_i denotes the parameter set. A larger $\phi_i(e; \Theta_i)$ indicates that entity e locates in the denser area of the embedding space and is more representative.

Considering that the query strategy scores are on incomparable scales, we convert them into percentiles as in [34]. Denote $\mathcal{P}_\phi(e, \mathcal{U})$ as the percentile of the score of e among the unlabeled data $\mathcal{U}$ in terms of query strategy ϕ. Accordingly, the converted percentile scores of degree centrality, PageRank centrality, and information density are denoted as $\mathcal{P}_{deg}$, $\mathcal{P}_{pr}$, and $\mathcal{P}_i$, respectively.

---

[2] We do not employ the uncertainty-based query strategies that are frequently used in the classification problems [7, 9], since it is complicated to characterize the uncertainty in the ranking-based problem setting [20]. We will investigate it in the future.

7.3.2 Reinforced Active Entity Selection via MAB

We leverage aforementioned query strategies to select the most informative entities for labeling. Considering that the significance of query strategies might vary in different iterations and no single query strategy can satisfy the need of all datasets, we propose to adaptively blend these strategies by adopting the multi-armed bandit (MAB) mechanism [26]. The MAB problems are some of the simplest reinforcement learning (RL) problems to solve, where we are given a slot machine with n arms (bandits) and each arm has its own probability distribution of success. Pulling any one of the arms yields a stochastic reward and the objective is to pull the arms such that the total reward collected in the long run can be maximized. Based on MAB, we treat each query strategy as an arm and approximate the importance of each strategy by estimating the expected reward (i.e., utility) of the corresponding arm. In this chapter, we adopt an extended framework of MAB, i.e., combinatorial MAB (CMAB) [6], which allows to play multiple arms in each iteration. Next, we elaborate the implementation details with regard to the alignment task.

Let Φ be the set of arms. In each iteration t, based on the percentile scores, each arm $\phi \in \Phi$ suggests its own set of entities to be labeled $Q_t(\phi)$, while the actual set of queried entities Q_t are chosen based on the utility score ε assigned to each unlabeled entity $e \in \mathcal{U}$, which is defined as:

$$\varepsilon_t(e) = \sum_{\phi}^{\Phi} \varepsilon_t(\phi) \mathcal{P}_{\phi}(e), \tag{7.2}$$

where $\varepsilon_t(\phi)$ is the utility score of arm ϕ in iteration t and $\mathcal{P}_{\phi}(e)$ is the percentile score of entity e in terms of query strategy ϕ. Then, the top b entities from the unlabeled entity set with the highest $\varepsilon_t(e)$ are selected as Q_t for querying the oracle in iteration t.

We estimate the utility of arm $\varepsilon(\phi)$ by taking into account the exploitation-exploration trade-off dilemma in MAB [6]. That is, we consider both the *exploitation* of the arm that has the highest expected payoff and the *exploration* to get more information about the expected payoffs of the other arms. Regarding the former, we define the expected reward of choosing arm ϕ in iteration t as the averaged reward it received from previous rounds:

$$\bar{\varepsilon}_t(\phi) = \frac{1}{t-1} \sum_{i=1}^{t-1} \hat{\varepsilon}_i(\phi), \tag{7.3}$$

where $\hat{\varepsilon}_i(\phi)$ is the reward received by arm ϕ in round i, which is defined as the change of alignment result on validation set:

$$\hat{\varepsilon}_i(\phi) = (F(\mathcal{L}_i \cup \mathcal{L}_i(\phi)) - F(\mathcal{L}_i)) + \frac{|\mathcal{L}_i(\phi)|}{b}(F(\mathcal{L}_i \cup Q_i) - F(\mathcal{L}_i)), \tag{7.4}$$

where $F(\cdot)$ denotes the value of a specific alignment measure (e.g., Hits@1, to be detailed in Sect. 7.5.1) on the validation set, which is generated by using the labeled data in the bracket. $\mathcal{L}_i$ represents the already labeled entities in iteration i. $\mathcal{L}_i(\phi) = Q_i(\phi) \cap Q_i$, denoting the set of *labeled* entities suggested by query strategy ϕ in iteration i. The difference between $F(\mathcal{L}_i \cup \mathcal{L}_i(\phi))$ and $F(\mathcal{L}_i)$ represents the *direct* change of alignment performance brought by arm ϕ. Besides, we reckon that the performance change caused by adding all the labeled entities Q_i can also be used to measure the utility of each arm. Hence, we use $\frac{|\mathcal{L}_i(\phi)|}{b}$ to denote the contribution of arm ϕ and multiply it with the overall performance change to produce the *implicit* change of alignment performance brought by arm ϕ.

Next, we move on to *exploration*. Following [6], to encourage leveraging the under-explored arms, we obtain the utility score of arm ϕ in iteration t by adjusting $\bar{\varepsilon}_t(\phi)$:

$$\varepsilon_t(\phi) = \bar{\varepsilon}_t(\phi) + \sqrt{\frac{3 \ln t}{2n_\phi}}, \tag{7.5}$$

where n_ϕ represents the total number of *labeled* entities suggested by arm ϕ until iteration t. As thus, the utility $\varepsilon(\phi)$ of arm ϕ is estimated by considering both the *exploitation* and *exploration*, which can provide more accurate signals for suggesting the entities to be labeled. Note that t starts from 1, and when $t = 1$, we omit the calculations of Eqs. (7.3)–(7.5) and set $\varepsilon_1(\phi)$ to 1 for all arms.

7.4   Contrastive Embedding Learning

Given the scarce labeled data generated by reinforced AL, in this section, we introduce contrastive entity representation learning that further mitigates the scarce supervision issue by mining supervision signals from the unlabeled data. We first introduce the semi-supervised alignment loss, the core of EA models. Then we introduce the graph encoders. Finally, we elaborate the unsupervised contrastive loss, as well as the training and inference processes.

7.4.1   Semi-supervised Alignment Loss

Since the entities (nodes) from different KGs cannot be compared directly, following current EA solutions, we first learn the entity structural embeddings of source and target KGs independently, i.e., O_s and O_t, and then devise a semi-supervised loss function to enforce the distance between the embeddings of the entities in the labeled entity pairs to be small and meanwhile the negative samples (i.e., nonequivalent entity pairs) to be large. Formally:

$$\mathcal{L}_s = \sum_{(u,v)\in\mathcal{S}} \sum_{(u',v')\in\mathcal{S}'_{(u,v)}} [dis(\boldsymbol{u},\boldsymbol{v}) + \gamma - dis(\boldsymbol{u}',\boldsymbol{v}')]_+, \qquad (7.6)$$

where $[\cdot]_+ = \max\{0,\cdot\}$, (u,v) is a labeled entity pair from the training data and $\mathcal{S}'_{(u,v)}$ represents the set of negative entity pairs obtained by corrupting (u,v) using nearest neighbor sampling [15]. $\boldsymbol{u}$ and $\boldsymbol{v}$ represent the embeddings of source and target entities retrieved from $\boldsymbol{O}_s$ and $\boldsymbol{O}_t$, respectively. $dis(\cdot,\cdot)$ is the distance function that measures the distance between two embeddings. γ is a hyper-parameter separating positive samples from negative ones.

In this chapter, the entity representation is obtained by aggregating the embeddings generated by two graph encoders: $\boldsymbol{O}_\omega = agg(\boldsymbol{Z}_\omega^{\psi_1}, \boldsymbol{Z}_\omega^{\psi_2})$, where agg is the aggregation function, which can be implemented as weighted average, concatenation, etc. $\boldsymbol{Z}_\omega^{\psi_1}$ and $\boldsymbol{Z}_\omega^{\psi_2}$ represent the embeddings generated from two different views. $\omega \in \{s,t\}$ denotes the source and target KG, respectively.

Note that we devise two graph encoders since (1) they can capture different views of the structural information and the integrated embeddings could be more expressive and (2) by devising a contrastive objective to enforce the embeddings of each entity in the two different views to agree with each other and meanwhile to be distinguished from the embeddings of other entities, the rich unlabeled data can be leveraged as supervision signals to learn discriminative entity representations and benefit the alignment.

7.4.2   Graph Encoders

In this chapter, we use two basic models, graph convolutional network (GCN) [13] and approximate personalized propagation of neural predictions [14], to capture the close and distant structural information of entities and generate different views of KG embeddings.[3]

The GCN model has been leveraged to generate entity embeddings by many previous works [30, 33]. It is a simple message passing algorithm. The inputs include the feature matrix of nodes X and the adjacency matrix of graph A. In the case of two message passing layers, the equation of GCN can be formulated as:

$$\boldsymbol{Z}^{\psi_1} = \mathrm{ReLU}\left(\hat{\boldsymbol{A}}\,\mathrm{ReLU}\left(\hat{\boldsymbol{A}}\boldsymbol{X}\boldsymbol{W}_0\right)\boldsymbol{W}_1\right), \qquad (7.7)$$

---

[3] Note that it is feasible to use other graph encoders here, e.g., the RREA embedding learning model (to be detailed in Sect. 7.5.2). We use two simple models to give prominence to the effects of reinforced AL and unsupervised CL strategies on EA.

where $\mathbf{Z}^{\psi_1}$ is the output entity embedding matrix. $\hat{A}$ is the symmetrically normalized adjacency matrix with self-loops. ReLU is the activation function, and $\mathbf{W}_0$ and $\mathbf{W}_1$ are the weight matrices.

While many approaches adopt GCN to learn entity representations, it is pointed out in [17] that when increasing the number of GCN layers, the alignment performance actually drops due to the oversmoothing issue. Therefore, we exploit the approximate personalized propagation [14] to generate the entity embeddings:

$$\mathbf{Z}^{(i)} = (1 - \alpha)\hat{A}\mathbf{Z}^{(i-1)} + \alpha\mathbf{X}, \quad i = 1, 2, \ldots, k, \tag{7.8}$$

where α is the teleport probability and k denotes the round of iterations. $\mathbf{Z}^{(0)} = \mathbf{X}$, and the initial feature matrix $\mathbf{X}$ acts as both the starting vector and the teleport set. $\mathbf{Z}^{\psi_2}$ is the output entity embedding matrix. Note that we remove the neural prediction network f_θ in the original model since it is not required in EA. We denote the resultant model as APP. By removing the weight matrices and nonlinearity in GCN, APP can capture distant structural information while retaining the quality of entity embeddings [14].

7.4.3 Unsupervised Contrastive Loss

Inspired by the successful application of contrastive learning (CL) on unsupervised graph representation learning [27, 36], in this chapter, we also devise a contrastive objective to distinguish the embeddings of the same entity under the two views from the embeddings of other entities, so as to leverage the supervision signals in the unlabeled data. Given an entity x_i, we denote its embedding generated by the first view as $\mathbf{Z}_\omega^{\psi_1}(i)$, and the embedding generated by the second view as $\mathbf{Z}_\omega^{\psi_2}(i)$, where $\omega \in \{s, t\}$ refers to the source and target KGs. These two embeddings form a positive sample. We consider the pairs of embeddings that contain $\mathbf{Z}_\omega^{\psi_1}(i)$ (or $\mathbf{Z}_\omega^{\psi_2}(i)$) and the embedding of another entity as the negative samples. Then, the contrastive object of the entity in the first view is defined as:

$$\ell_\omega^{\psi_1}(x_i) = -\log \frac{e^{\theta\left(\mathbf{Z}_\omega^{\psi_1}(i), \mathbf{Z}_\omega^{\psi_2}(i)\right)}}{e^{\theta\left(\mathbf{Z}_\omega^{\psi_1}(i), \mathbf{Z}_\omega^{\psi_2}(i)\right)} + \mathcal{N}_{cross} + \mathcal{N}_{intra}}, \tag{7.9}$$

$$\mathcal{N}_{cross} = \sum_{k=1}^{n_\omega} \mathbf{1}_{[k \neq i]} e^{\theta\left(\mathbf{Z}_\omega^{\psi_1}(i), \mathbf{Z}_\omega^{\psi_2}(k)\right)} \tag{7.10}$$

$$\mathcal{N}_{intra} = \sum_{k=1}^{n_\omega} \mathbf{1}_{[k \neq i]} e^{\theta\left(\mathbf{Z}_\omega^{\psi_1}(i), \mathbf{Z}_\omega^{\psi_1}(k)\right)} \tag{7.11}$$

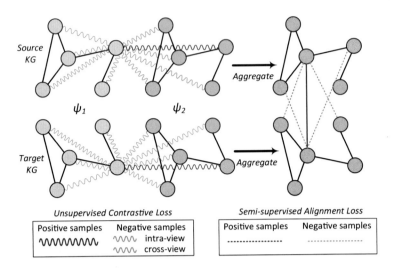

Fig. 7.3 Illustration of the losses

where $\theta(\cdot, \cdot)$ is a score function that calculates the similarity between two embeddings, which is implemented as $\theta(\cdot, \cdot) = f(g(\cdot), g(\cdot))$, where $g(\cdot)$ is a multilayer perceptron (MLP) with nonlinear activation functions for transforming the embeddings, and $f(\cdot, \cdot)$ is a similarity metric capturing the similarity between embeddings. $\mathbf{1}_{[\cdot]}$ is an indicator function which equals to 1 if the argument inside the bracket holds, and 0 otherwise. n_ω is the number of entities in the KG. In the denominator, the first term is the positive sample, the second term $\mathcal{N}_{cross}$ corresponds to the cross-view negative samples, and the third term $\mathcal{N}_{intra}$ corresponds to the intra-view negative samples. Detailed illustrations can be found in Fig. 7.3. The contrastive object of the second view $\ell_\omega^{\psi_2}(x_i)$ is defined similarly. As thus, the overall unsupervised loss is defined as:

$$\mathcal{L}_u = \frac{1}{2n_s} \sum_{i=1}^{n_s} \left[\ell_s^{\psi_1}(e_i) + \ell_s^{\psi_2}(e_i) \right] + \frac{1}{2n_t} \sum_{i=1}^{n_t} \left[\ell_t^{\psi_1}(e_i) + \ell_t^{\psi_2}(e_i) \right], \quad (7.12)$$

where n_s and n_t denote the number of entities in the source and target KGs, respectively.

Model Training Finally, we combine the semi-supervised alignment loss and the unsupervised contrastive loss, resulting in the loss function of our proposed model:

$$\mathcal{L} = \mathcal{L}_s + \lambda_u \mathcal{L}_u, \quad (7.13)$$

where $\lambda_u > 0$ is the hyper-parameter balancing the two objectives.

7.4.4 Alignment Inference

After obtaining the learned unified embeddings, the alignment results can thus be inferred. For each source entity, we calculate its distance with all target entities according to a specific distance metric and consider the entity with the smallest distance as the match. We describe the overall procedure of RAC in Algorithm 1.

Algorithm 1: Reinforced active entity alignment

Input : $\mathcal{G}_s$ and $\mathcal{G}_t$: source and target KGs; S: labeled data; B: labeling budget; An oracle to label entities.
Output : $\mathcal{A}$: the set of aligned entity pairs.
1 $S' \leftarrow S$;
2 **while** $|S'| - |S| < B$ **do**
3   Generate query strategies (using entity representations);
4   Use MAB to select b entities for the oracle to label and obtain the labeled entity set S_b;
5      $S' \leftarrow S' \cup S_b$;
6   Generate entity representations using graph encoders;
7   Calculate $\mathcal{L}_s$ using S' via Eq. (7.6);
8   Calculate $\mathcal{L}_u$ using Eqs. (7.9) and (7.12);
9   Calculate $\mathcal{L}$ for training unified entity representations;
10 Infer the results $\mathcal{A}$ using the learned representations;
11 **return** $\mathcal{A}$;

7.5 Experiment

In this section, we empirically evaluate our proposed model[4] by answering the following questions:

- **RQ1**: Does RAC outperform baseline alignment models on EA with limited supervision? Are the contrastive learning and reinforced AL modules useful?
- **RQ2**: Where does the performance gain brought by CL come from? Is the combination of embeddings from different views effective enough? Is it sensitive to hyper-parameters?
- **RQ3**: Can the reinforced AL strategy be applied on baseline models? Is it better than using the query strategies separately or blending strategies with equal weights?

---

[4] The source code is available at https://github.com/DexterZeng/RAC.

Table 7.1 Statistics of the datasets used for evaluation

Dataset	KG pairs	#Triples	#Ents	#Rels	#Aligns
DBP15K$_{ZH-EN}$	DBpedia (ZH&EN)	165,556	38,960	3,024	15,000
DBP15K$_{JA-EN}$	DBpedia (JA&EN)	170,698	39,594	2,452	15,000
DBP15K$_{FR-EN}$	DBpedia (FR&EN)	221,720	39,654	2,111	15,000
SRPRS$_{EN-FR}$	DBpedia (EN&FR)	70,040	30,000	398	15,000
SRPRS$_{EN-DE}$	DBpedia (EN&DE)	75,740	30,000	342	15,000
SRPRS$_{DBP-WD}$	DBpedia & Wikidata	78,580	30,000	397	15,000
SRPRS$_{DBP-YG}$	DBpedia & YAGO	70,317	30,000	253	15,000
DBP-FB	DBpedia & Freebase	208,388	55,403	1,289	25,542

EN, ZH, JA, FR, and DE refer to the English, Chinese, Japanese, French, and German version of DBpedia, respectively #Triples, #Ents, #Rels, and #Aligns denote the number of triples, entities, relations, and gold alignment data in each dataset, respectively

7.5.1 Experimental Settings

Datasets Following previous works, we adopt three popular EA datasets for evaluation: (1) DBP15K [23], which includes three cross-lingual KG pairs extracted from DBpedia; (2) SRPRS [11], which comprises two cross-lingual and two mono-lingual KG pairs extracted from DBpedia, Wikidata, and YAGO; and (3) DBP-FB [35], which is a mono-lingual KG pair extracted from DBpedia and Freebase. In each KG pair, 70%, 10%, and 20% of the gold standards are used for testing, validation, and training, respectively. Since we study EA with limited supervision, we only keep 500 seed entity pairs as the initial training set. Then, according to the labeling budget, we select the entities from the rest of the training data for annotation and add the labeled entity pairs into the initial training set. The details of datasets can be found in Table 7.1.

Implementation Details Regarding the *query strategies*, we set the damping parameter in Eq. (7.1) to the default value 0.85. We set b, the number of entities selected in each iteration, to 50. As to the *semi-supervised alignment loss* in Eq. (7.6), we adopt the Euclidean distance as $dis(\cdot, \cdot)$ and select γ among $[1, 3, 5, 10]$. We implement the embedding aggregation function as: $agg(\mathbf{Z}_\omega^{\psi_1}, \mathbf{Z}_\omega^{\psi_2}) = \lambda_e \mathbf{Z}_\omega^{\psi_1} + (1 - \lambda_e)\mathbf{Z}_\omega^{\psi_2}$, where $\lambda_e \in (0, 1)$ is the hyper-parameter that balances the weights of the two views, and we select it among $[0.2, 0.4, 0.6, 0.8]$. As for the *graph encoders*, we follow previous works [30, 33] by adopting two two-layer GCNs. We follow the original work of APP [14] and directly set the teleport probability α in Eq. (7.8) to 0.2, and the round of propagations k to 5. The dimensionality of entity embeddings is set to 100. Concerning the *unsupervised contrastive loss* in Eq. (7.9), we implement $g(\cdot)$ as two-layer MLP with a nonlinear activation *elu* and adopt the cosine similarity as $f(\cdot, \cdot)$. We select λ_u in Eq. (7.13) among $[0.05, 0.1, 0.15, 0.2, 0.25, 0.3, 0.35]$ and adopt Adam optimizer to minimize the training objective. The distance function in the alignment inference process is set to the Euclidean distance.

By tuning the hyper-parameters on the validation set, we set γ to 1, λ_e to 0.2, and λ_u to 0.2. The experiments are conducted on a personal computer with the Ubuntu system, an Intel Core i7-4790 CPU, an NVIDIA GeForce GTX TITAN X GPU, and a 32 GB memory. We conduct the experiments for five independent runs and report the averaged performance (and the standard deviation) on each dataset.

Evaluation Metrics Following the convention [5], for each source entity in the test set, we rank the target entities ascendingly according to the embedding distance as in Sect. 7.4.4 and adopt Hits@1 as the evaluation metric, which is defined as the percentage of source entities whose ground-truth target entity is ranked first. Note that the Hits@1 results are represented in percentages, and the bold figures in the tables represent the best results.

Competing Methods The majority of state-of-the-art EA methods focus on designing advanced representation learning models to capture more useful structural information for alignment. In comparison, our proposed framework RAC aims to improve the alignment performance under limited supervision by using reinforced AL and CL, which are agnostic to the choices of these embedding learning models. RAC can be applied on these methods to improve their capability of dealing with scare supervision signals. Hence, the main goal of this chapter is not to compare with these state-of-the-art models, but with the methods that improve EA performance under scarce supervision. In this light, we compare RAC with a very recent work [2], ALEA, which harnessed AL for EA. Specifically, we adopt the most performant variants of ALEA, i.e., ALEA-D and ALEA-B, as the baseline models, which leverage the degree and betweenness centrality as the query strategies, respectively.

Noteworthily, to demonstrate the wide applicability of RAC, we employ it on the most performant embedding learning model RREA [17], as well as a state-of-the-art EA model that leverages auxiliary information, CEA [33], in Sect. 7.5.2.

7.5.2 Main Results (RQ1)

We report the alignment results in Table 7.2 by setting the labeling budget B to 500 and 1500, respectively. It can be observed that RAC significantly outperforms the embedding learning-based baseline models GCN and APP across all datasets (over 40% on DBP-FB), showcasing the effectiveness of our proposal when the supervision signals are limited. Particularly, RAC ($B = 500$) even achieves comparable results to GCN ($B = 1500$) on SRPRS$_{\text{EN-FR}}$ and DBP-FB, which validates that, to reach a certain performance target, adopting RAC significantly reduces manual labeling effort. Besides, it is notable that the improvement is more prominent when there are fewer labeled data. For instance, RAC outperforms APP by over 15% on most datasets when $B = 500$, while the improvement is less than 15% on most datasets when $B = 1500$.

Table 7.2 Hits@1 results under different budgets

Budget	Method	DBP15K			SRPRS				DBP-FB
		DBP15K$_{ZH-EN}$	DBP15K$_{JA-EN}$	DBP15K$_{FR-EN}$	SRPRS$_{EN-FR}$	SRPRS$_{EN-DE}$	SRPRS$_{DBP-WD}$	SRPRS$_{DBP-YG}$	
$B=500$	GCN	22.53 ± 0.41	20.93 ± 0.39	22.95 ± 0.48	14.73 ± 0.37	30.58 ± 0.27	15.59 ± 0.18	21.64 ± 0.21	5.97 ± 0.17
	APP	24.65 ± 0.27	24.25 ± 0.48	25.78 ± 0.65	16.96 ± 0.33	31.60 ± 0.30	17.35 ± 0.24	23.40 ± 0.41	6.78 ± 0.12
	ALEA-D	27.41 ± 0.10	25.89 ± 0.14	26.94 ± 0.08	18.88 ± 0.20	32.47 ± 0.12	19.67 ± 0.10	25.34 ± 0.19	9.01 ± 0.07
	ALEA-B	26.03 ± 0.19	24.95 ± 0.21	26.58 ± 0.18	17.84 ± 0.10	32.35 ± 0.19	18.64 ± 0.17	25.30 ± 0.14	8.56 ± 0.11
	RAC	**28.64 ± 0.33**	**28.71 ± 0.56**	**29.44 ± 0.18**	**19.94 ± 0.21**	**34.05 ± 0.30**	**20.70 ± 0.30**	**27.29 ± 0.17**	**9.44 ± 0.12**
$B=1500$	GCN	31.48 ± 0.40	32.14 ± 0.56	32.51 ± 0.35	20.48 ± 0.31	35.64 ± 0.18	22.35 ± 0.38	27.62 ± 0.33	9.99 ± 0.14
	APP	33.08 ± 0.42	35.26 ± 0.40	34.79 ± 0.36	22.03 ± 0.33	36.47 ± 0.09	23.49 ± 0.20	28.76 ± 0.32	10.54 ± 0.16
	ALEA-D	35.41 ± 0.17	36.77 ± 0.14	35.96 ± 0.13	23.65 ± 0.12	37.19 ± 0.07	24.76 ± 0.09	30.52 ± 0.08	13.70 ± 0.08
	ALEA-B	34.17 ± 0.25	35.86 ± 0.17	35.72 ± 0.13	23.18 ± 0.06	36.49 ± 0.14	24.72 ± 0.14	30.08 ± 0.10	11.25 ± 0.07
	RAC	**37.65 ± 0.13**	**39.51 ± 0.16**	**38.45 ± 0.28**	**25.48 ± 0.18**	**39.27 ± 0.37**	**26.72 ± 0.19**	**32.68 ± 0.14**	**15.35 ± 0.36**

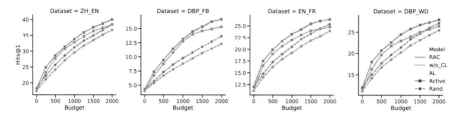

Fig. 7.4   Hits@1 results of ablation study. The shaded area denotes the standard deviation

Then, by comparing the results of RAC with ALEA-D and ALEA-B, it is obvious that our proposed model is more effective and robust than existing AL-based EA models given scarce labeled data. The superior performance can be attributed to the reinforced AL strategy and the contrastive representation learning, which we will analyze in detail in the following.

Ablation Results   To examine the usefulness of the two key components—the unsupervised contrastive loss and the reinforced AL strategy—in RAC, we conduct ablation study. As shown in Fig. 7.4, we select the labeling budget B among [250, 500, 750, 1000, 1250, 1500, 1750, 2000] and obtain the corresponding alignment results of RAC -Active, RAC -Rand., RAC w/o CL -Active, and RAC w/o CL -Rand., where -Active denotes using our proposed reinforced AL strategy, -Rand. denotes selecting the entities randomly, and w/o CL denotes removing the unsupervised contrastive loss. Note that, in the interest of space, we only select representative KG pairs from each dataset and report their results, among which $DBP15K_{ZH-EN}$ and $SRPRS_{EN-FR}$ are cross-lingual datasets, while $SRPRS_{DBP-WD}$ and $DBP-FB$ are mono-lingual ones.

It reads from Fig. 7.4 that the reinforced AL and CL strategies both contribute positively to the overall performance. More concretely, with the increase of labeling budget, the effectiveness of the AL strategy becomes less significant, while the unsupervised contrastive learning loss begins to play a more important role. This could be ascribed to the fact that: (1) the quality of the entities selected by AL drops with the increase of budget since the valuable entities have already been chosen in the early stages and (2) the effectiveness of CL relies on the quality of entity representations, which is improved when there are more labeled data.

Applying RAC on Embedding Learning-Based EA Model   We apply RAC on RREA, the most performant EA method so far, to see whether RAC would improve its capability of dealing with limited supervision. Specifically, we follow the implementation details in the original paper [17] and contrast the entity embeddings learned by it with the embeddings generated by GCN and then conduct the reinforced AL. The results are provided in Table 7.3, which validate that RAC can be applied on existing EA models to improve their performance under limited supervision, and the improvement is more notable when there are fewer labeled data ($B = 500$ vs. $B = 1500$).

Table 7.3 Hits@1 results of applying RAC on RREA and CEA

B	Method	DBP15K$_{ZH-EN}$	SRPRS$_{EN-FR}$	SRPRS$_{DBP-WD}$	DBP-FB
500	RREA	48.42 ± 0.38	32.07 ± 0.41	35.14 ± 0.29	17.11 ± 0.25
	+RAC	$\mathbf{51.84 \pm 0.18}$	$\mathbf{35.58 \pm 0.41}$	$\mathbf{38.04 \pm 0.22}$	$\mathbf{21.84 \pm 0.26}$
1500	RREA	57.63 ± 0.32	37.29 ± 0.36	41.74 ± 0.11	22.65 ± 0.20
	+RAC	$\mathbf{59.74 \pm 0.36}$	$\mathbf{39.82 \pm 0.29}$	$\mathbf{43.66 \pm 0.20}$	$\mathbf{27.24 \pm 0.21}$
B	Method	DBP15K$_{ZH-EN}$	SRPRS$_{EN-FR}$	DBP15K$_{JA-EN}$	SRPRS$_{EN-DE}$
500	CEA	64.19 ± 0.12	90.08 ± 0.22	71.57 ± 0.43	91.09 ± 0.52
	+RAC	$\mathbf{68.18 \pm 0.27}$	$\mathbf{90.84 \pm 0.45}$	$\mathbf{73.60 \pm 0.34}$	$\mathbf{91.70 \pm 0.91}$
1500	CEA	68.37 ± 0.15	91.86 ± 0.27	75.58 ± 0.42	92.28 ± 0.51
	+RAC	$\mathbf{71.25 \pm 0.25}$	$\mathbf{92.03 \pm 0.21}$	$\mathbf{76.98 \pm 0.27}$	$\mathbf{93.18 \pm 0.24}$

Applying RAC on EA Model that Uses Auxiliary Information We apply RAC on CEA [33], an EA model leveraging the entity name information to complement KG structural information for alignment. The results are provided in Table 7.3, which demonstrate that our proposal is also effective on EA models harnessing auxiliary information. We notice that the improvements on SRPRS$_{EN-FR}$ and SRPRS$_{EN-DE}$ are not significant. This is because the entity name information in SRPRS can already provide very accurate alignment signals, e.g., solely comparing the entity names can lead to ground-truth performance on SRPRS$_{DBP-YG}$ and SRPRS$_{DBP-WD}$ [35] (and thus we omit their results in Table 7.3). This unveils that it is more beneficial to study EA with scarce supervision when the auxiliary information is not available or of low quality (as is often the case) [35].

7.5.3   Experiments on Contrastive Learning (RQ2)

In this subsection, we carefully examine the effectiveness of unsupervised CL. We first empirically validate that the main performance enhancement brought by CL comes from the unsupervised contrastive loss itself rather than the combination of embeddings. Then we conduct parameter analysis to show its robustness.

Comparison with *mere* Combination of Embeddings Since different representation learning models capture different structural information in KGs, one might wonder whether the effectiveness of unsupervised CL mainly comes from the combination of embeddings. To investigate this issue, we remove the effect of AL and report the results of GCN, APP, the combination of these two embeddings (denoted as Comb.), and the combination of these two embeddings with unsupervised CL loss (denoted as Comb.+CL) in Table 7.4. It shows that, compared with utilizing the representation learning models separately, Comb. only slightly improves the alignment performance in some cases and even brings down the results under a few settings, e.g., $B = 500$. After adding the unsupervised contrastive loss,

Table 7.4 Hits@1 results of variants of CL on DBP15K$_{\text{ZH-EN}}$ after removing the influence of AL

Method	$B = 500$	$B = 1000$	$B = 1500$	$B = 2000$
GCN	22.53 ± 0.41	27.65 ± 0.49	31.48 ± 0.40	34.93 ± 0.44
APP	24.65 ± 0.27	29.66 ± 0.22	33.08 ± 0.42	36.22 ± 0.11
Comb.	24.17 ± 0.34	29.61 ± 0.10	33.56 ± 0.41	36.79 ± 0.19
Comb.+CL	25.71 ± 0.26	31.23 ± 0.42	35.07 ± 0.45	$\mathbf{38.45 \pm 0.21}$
$\lambda_u = 0.05$	25.56 ± 0.31	30.60 ± 0.20	34.57 ± 0.50	37.59 ± 0.29
$\lambda_u = 0.10$	$\mathbf{25.81 \pm 0.21}$	31.12 ± 0.19	35.09 ± 0.62	38.13 ± 0.31
$\lambda_u = 0.15$	25.70 ± 0.25	$\mathbf{31.24 \pm 0.35}$	$\mathbf{35.24 \pm 0.69}$	38.39 ± 0.28
$\lambda_u = 0.25$	25.58 ± 0.38	31.11 ± 0.50	35.15 ± 0.63	38.27 ± 0.23
$\lambda_u = 0.30$	25.54 ± 0.29	31.03 ± 0.48	35.01 ± 0.57	38.23 ± 0.10
$\lambda_u = 0.35$	25.46 ± 0.30	30.89 ± 0.59	34.90 ± 0.59	38.08 ± 0.17

Note that Comb.+CL is equivalent to $\lambda_u = 0.2$ in the bottom half of the table

Comb.+CL achieves superior results than Comb. and APP across all settings. This demonstrates the significance of using CL to mine supervision signals from the abundant unlabeled data. Furthermore, by comparing RAC with RAC w/o CL (which combines the two representations) in Fig. 7.4, we can conclude that the contrastive loss is effective with or without AL.

Sensitivity Analysis We conduct sensitivity analysis on a critical hyper-parameter in the contrastive entity representation learning, λ_u, which determines the contributions of semi-supervised alignment loss $\mathcal{L}_s$ and unsupervised contrastive loss $\mathcal{L}_u$ to the overall training objective, to show the stability of the model under perturbation of the hyper-parameter. Since it is intuitive that $\mathcal{L}_s$ can provide more accurate signals for alignment compared with $\mathcal{L}_u$, we vary λ_u from 0.05 to 0.35 and report the results in Table 7.4. From the table, it can be observed that the alignment performance is relatively stable when λ_u is not too large. We thus conclude that, overall, our model is robust to the perturbation of λ_u.

7.5.4 Experiments on Reinforced AL (RQ3)

In this subsection, we aim to examine the usefulness of the reinforced AL component. We first demonstrate that the reinforced AL strategy can be applied on the baseline models to improve their performance given scarce labeled data. Next, we empirically verify that using our proposed reinforced AL to blend query strategies can lead to better results than using these strategies individually or combining the query strategies with equal weights.

Effectiveness of Reinforced AL on Baseline Models We apply our proposed reinforced AL on the baseline models and report the results in Fig. 7.5. It shows that

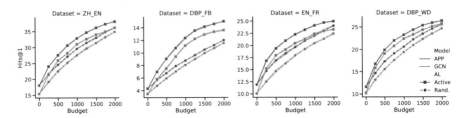

Fig. 7.5 Hits@1 results of applying AL on baseline models. The shaded area denotes the standard deviation

the performance of both APP and GCN is enhanced after applying the reinforced AL strategy, and the improvement is more prominent when the budget is smaller.

Comparison with Using Query Strategies Individually To verify that blending the query strategies with MAB is more effective than using these strategies individually, we replace reinforced AL in RAC with degree centrality, PageRank centrality, and information density, resulting in RAC-Deg, RAC-Pr, and RAC-Emb, respectively, and report the results in Table 7.5. It shows that, overall speaking, the reinforced active entity selection strategy can lead to better alignment results than using query strategies individually.

Comparison with Combination with Equal Weights To demonstrate that reinforced AL can adaptively integrate query strategies and lead to better alignment performance, we compare it with blending query strategies with equal weights (RAC-Avg) and provide the results in Table 7.5. It can be observed that RAC is more effective than RAC-Avg, especially when the budget value is small, showcasing the importance of combining query strategies adaptively.

7.6 Related Work

Entity Alignment The task of EA has been intensively studied over the last few years [35]. The majority of existing EA literature [4, 5, 10, 23, 30] are devoted to learning better entity representations using the KG embedding techniques such as TransE and GCN. Specifically, some propose to capture the neighboring information [11, 25] for learning expressive entity representations, while some propose to model the relations to help guide the alignment of entities [24, 31]. All of these approaches require seed entity pairs to project entity embeddings from different KGs into a unified space, where the entities can be directly compared across KGs. Nevertheless, such labeled data are hard to obtain in real-life settings. To reduce the reliance on labeled data, some efforts are devoted to aligning entities in unsupervised settings [32]. They leverage the auxiliary (side information) of KGs, such as attributes and entity names, to produce the pseudo-labeled data, which are then used to learn the unified structural embeddings. Nevertheless, the effectiveness

Table 7.5 Hits@1 results of the variants of reinforced AL

Budget	Method	DBP15K			SRPRS				DBP-FB	AVG
		DBP15K$_{ZH-EN}$	DBP15K$_{JA-EN}$	DBP15K$_{FR-EN}$	SRPRS$_{EN-FR}$	SRPRS$_{EN-DE}$	SRPRS$_{DBP-WD}$	SRPRS$_{DBP-YG}$		
$B = 500$	RAC	**28.64 ± 0.33**	**28.71 ± 0.56**	**29.44 ± 0.18**	**19.94 ± 0.21**	**34.05 ± 0.30**	**20.70 ± 0.30**	**27.29 ± 0.17**	9.44 ± 0.12	**24.78**
	RAC-Deg	28.61 ± 0.19	27.20 ± 0.18	27.36 ± 0.22	19.73 ± 0.16	33.57 ± 0.15	20.66 ± 0.18	27.03 ± 0.27	**9.67 ± 0.19**	24.23
	RAC-Pr	28.05 ± 0.17	26.31 ± 0.30	28.73 ± 0.29	19.75 ± 0.12	33.92 ± 0.19	20.33 ± 0.14	27.27 ± 0.11	9.18 ± 0.18	24.19
	RAC-Emb	24.85 ± 0.30	23.84 ± 0.14	25.61 ± 0.54	16.59 ± 0.46	31.33 ± 0.30	16.78 ± 0.43	22.85 ± 0.40	5.85 ± 0.29	20.96
	RAC-Avg	28.48 ± 0.28	28.28 ± 0.33	29.35 ± 0.18	19.72 ± 0.16	34.00 ± 0.51	20.12 ± 0.25	27.20 ± 0.17	9.09 ± 0.20	24.53
$B = 1500$	RAC	**37.65 ± 0.13**	**39.51 ± 0.16**	**38.45 ± 0.28**	**25.48 ± 0.18**	39.27 ± 0.37	**26.72 ± 0.19**	**32.68 ± 0.14**	**15.35 ± 0.36**	**31.89**
	RAC-Deg	37.17 ± 0.41	38.17 ± 0.25	37.25 ± 0.17	24.95 ± 0.27	38.88 ± 0.12	26.40 ± 0.21	32.18 ± 0.23	15.00 ± 0.10	31.25
	RAC-Pr	36.68 ± 0.11	38.63 ± 0.19	37.82 ± 0.19	24.91 ± 0.22	**39.35 ± 0.14**	26.52 ± 0.20	32.63 ± 0.11	14.94 ± 0.13	31.43
	RAC-Emb	34.78 ± 0.19	37.87 ± 0.24	37.31 ± 0.22	22.93 ± 0.36	38.01 ± 0.14	24.16 ± 0.18	29.47 ± 0.32	9.60 ± 0.26	29.27
	RAC-Avg	37.64 ± 0.15	39.44 ± 0.14	38.37 ± 0.13	25.33 ± 0.17	39.20 ± 0.10	26.70 ± 0.10	32.60 ± 0.17	15.33 ± 0.26	31.83

of these approaches is largely restrained by the quality of side information. In practice, the auxiliary information could be unavailable or unevenly distributed [35].

EA with Limited Supervision The most similar work to ours is [2], which examines the effectiveness of various heuristics from AL in terms of improving EA performance under limited supervision. Our work differs from [2] in that (1) we devise a reinforced AL framework to adaptively blend query strategy heuristics and (2) we exploit the idea of CL to help further improve the EA performance. We also empirically validate the superiority of our proposal over [2].

Reinforced Active Learning Reinforced AL approaches have been developed also for other related problems, where RL is used to take the role of traditional query strategy heuristics [7–9]. To tackle cross-lingual named entity recognition task, Fang et al. design a deep Q-network to select data for annotation in a streaming setting [8]. In [7, 9], different multi-armed bandit models [6] are used to learn active discriminative network representations for the node classification task. Note that the MAB mechanism implemented in RAC differs from theirs and is developed specifically for the alignment task.

Contrastive Learning on Graphs Recently, contrastive learning (CL) has emerged as a successful method for unsupervised graph representation learning [27, 36]. CL is an active field of self-supervised learning, which can generate data representations by learning to encode the similarities or dissimilarities among a set of unlabeled examples [12]. The intuition behind is that the rich unlabeled data themselves can be used as supervision signals to help guide model training. In this chapter, we also exploit this idea and leverage the abundant unlabeled entities to facilitate the alignment.

7.7 Conclusion

State-of-the-art EA approaches are overly dependent on labeled data, which are difficult to obtain in practical settings. In response, we propose a reinforced active framework RAC to tackle EA with scarce supervision. In each labeling iteration, RAC selects the valuable entities to be labeled according to the multi-armed bandit mechanism that blends different query strategies. Then, given the limited labeled data, it mines useful supervision signals from the rich unlabeled data to help generate more accurate entity representations (and the alignment results). We evaluate RAC on popular EA benchmarks, and the empirical results validate that RAC is effective at coping with limited labeled data. Besides, we also demonstrate that RAC is a general framework to tackle EA with scarce supervision and can be employed on top of existing EA solutions.

References

1. S. Auer, C. Bizer, G. Kobilarov, J. Lehmann, R. Cyganiak, and Z. G. Ives. Dbpedia: A nucleus for a web of open data. In *ISWC*, pages 722–735, 2007.
2. M. Berrendorf, E. Faerman, and V. Tresp. Active learning for entity alignment. In *ECIR*, pages 48–62, 2021.
3. K. D. Bollacker, C. Evans, P. Paritosh, T. Sturge, and J. Taylor. Freebase: a collaboratively created graph database for structuring human knowledge. In *SIGMOD*, pages 1247–1250, 2008.
4. J. Chen, Z. Li, P. Zhao, A. Liu, L. Zhao, Z. Chen, and X. Zhang. Learning short-term differences and long-term dependencies for entity alignment. In *ISWC*, volume 12506, pages 92–109, 2020.
5. M. Chen, Y. Tian, M. Yang, and C. Zaniolo. Multilingual knowledge graph embeddings for cross-lingual knowledge alignment. In *IJCAI*, pages 1511–1517, 2017.
6. W. Chen, Y. Wang, and Y. Yuan. Combinatorial multi-armed bandit: General framework and applications. In *ICML*, volume 28, pages 151–159, 2013.
7. X. Chen, G. Yu, J. Wang, C. Domeniconi, Z. Li, and X. Zhang. Activehne: Active heterogeneous network embedding. In *IJCAI*, pages 2123–2129, 2019.
8. M. Fang, Y. Li, and T. Cohn. Learning how to active learn: A deep reinforcement learning approach. In *EMNLP*, pages 595–605, 2017.
9. L. Gao, H. Yang, C. Zhou, J. Wu, S. Pan, and Y. Hu. Active discriminative network representation learning. In *IJCAI*, pages 2142–2148, 2018.
10. H. Guo, J. Tang, W. Zeng, X. Zhao, and L. Liu. Multi-modal entity alignment in hyperbolic space. *Neurocomputing*, 2021.
11. L. Guo, Z. Sun, and W. Hu. Learning to exploit long-term relational dependencies in knowledge graphs. In *ICML*, pages 2505–2514, 2019.
12. R. D. Hjelm, A. Fedorov, S. Lavoie-Marchildon, K. Grewal, P. Bachman, A. Trischler, and Y. Bengio. Learning deep representations by mutual information estimation and maximization. In *ICLR*, 2019.
13. T. N. Kipf and M. Welling. Semi-supervised classification with graph convolutional networks. In *ICLR*. OpenReview.net, 2017.
14. J. Klicpera, A. Bojchevski, and S. Günnemann. Predict then propagate: Graph neural networks meet personalized pagerank. In *ICLR*. OpenReview.net, 2019.
15. C. Li, Y. Cao, L. Hou, J. Shi, J. Li, and T. Chua. Semi-supervised entity alignment via joint knowledge embedding model and cross-graph model. In *EMNLP*, pages 2723–2732, 2019.
16. X. Mao, W. Wang, H. Xu, M. Lan, and Y. Wu. MRAEA: an efficient and robust entity alignment approach for cross-lingual knowledge graph. In *WSDM*, pages 420–428. ACM, 2020.
17. X. Mao, W. Wang, H. Xu, Y. Wu, and M. Lan. Relational reflection entity alignment. In *CIKM*, pages 1095–1104, 2020.
18. M. E. J. Newman. *Networks: An Introduction*. Oxford University Press, 2010.
19. H. Nie, X. Han, L. Sun, C. M. Wong, Q. Chen, S. Wu, and W. Zhang. Global structure and local semantics-preserved embeddings for entity alignment. In *IJCAI*, pages 3658–3664, 2020.
20. N. Ostapuk, J. Yang, and P. Cudré-Mauroux. Activelink: Deep active learning for link prediction in knowledge graphs. In *WWW*, pages 1398–1408, 2019.
21. B. Settles. Active learning literature survey. Technical report, 2010.
22. F. M. Suchanek, G. Kasneci, and G. Weikum. Yago: a core of semantic knowledge. In *WWW*, pages 697–706, 2007.
23. Z. Sun, W. Hu, and C. Li. Cross-lingual entity alignment via joint attribute-preserving embedding. In *ISWC*, pages 628–644, 2017.
24. Z. Sun, J. Huang, W. Hu, M. Chen, L. Guo, and Y. Qu. Transedge: Translating relation-contextualized embeddings for knowledge graphs. In *ISWC*, pages 612–629, 2019.
25. Z. Sun, C. Wang, W. Hu, M. Chen, J. Dai, W. Zhang, and Y. Qu. Knowledge graph alignment network with gated multi-hop neighborhood aggregation. In *AAAI*, pages 222–229, 2020.

26. R. S. Sutton and A. G. Barto. Reinforcement learning: An introduction. *IEEE Trans. Neural Networks*, 9(5):1054–1054, 1998.
27. P. Velickovic, W. Fedus, W. L. Hamilton, P. Liò, Y. Bengio, and R. D. Hjelm. Deep graph infomax. In *ICLR*, 2019.
28. D. Vrandecic and M. Krötzsch. Wikidata: a free collaborative knowledgebase. *Commun. ACM*, 57(10):78–85, 2014.
29. S. Wan, S. Pan, J. Yang, and C. Gong. Contrastive and generative graph convolutional networks for graph-based semi-supervised learning. In *AAAI*, pages 10049–10057, 2021.
30. Z. Wang, Q. Lv, X. Lan, and Y. Zhang. Cross-lingual knowledge graph alignment via graph convolutional networks. In *EMNLP*, pages 349–357, 2018.
31. Y. Wu, X. Liu, Y. Feng, Z. Wang, R. Yan, and D. Zhao. Relation-aware entity alignment for heterogeneous knowledge graphs. In *IJCAI*, pages 5278–5284, 2019.
32. W. Zeng, X. Zhao, J. Tang, X. Li, M. Luo, and Q. Zheng. Towards entity alignment in the open world: An unsupervised approach. In *DASFAA*, pages 272–289. Springer, 2021.
33. W. Zeng, X. Zhao, J. Tang, and X. Lin. Collective entity alignment via adaptive features. In *ICDE*, pages 1870–1873. IEEE, 2020.
34. Y. Zhang, M. Lease, and B. C. Wallace. Active discriminative text representation learning. In *AAAI*, pages 3386–3392, 2017.
35. X. Zhao, W. Zeng, J. Tang, W. Wang, and F. M. Suchanek. An experimental study of state-of-the-art entity alignment approaches. *IEEE Transactions on Knowledge and Data Engineering*, pages 1–1, 2020.
36. Y. Zhu, Y. Xu, F. Yu, Q. Liu, S. Wu, and L. Wang. Graph contrastive learning with adaptive augmentation. In *WWW*, pages 2069–2080, 2021.

Chapter 8
Unsupervised Entity Alignment

Abstract State-of-the-art entity alignment solutions tend to rely on labeled data for model training. Additionally, they work under the closed-domain setting and cannot deal with entities that are unmatchable. To address these deficiencies, we offer an unsupervised framework that performs entity alignment in the open world. Specifically, we first mine useful features from the side information of KGs. Then, we devise an unmatchable entity prediction module to filter out unmatchable entities and produce preliminary alignment results. These preliminary results are regarded as the pseudo-labeled data and forwarded to the progressive learning framework to generate structural representations, which are integrated with the side information to provide a more comprehensive view for alignment. Finally, the progressive learning framework gradually improves the quality of structural embeddings and enhances the alignment performance by enriching the pseudo-labeled data with alignment results from the previous round. Our solution does not require labeled data and can effectively filter out unmatchable entities. Comprehensive experimental evaluations validate its superiority.

8.1 Introduction

State-of-the-art EA solutions [2–5] assume that equivalent entities usually possess similar neighboring information. Consequently, they utilize KG embedding models, e.g., TransE [6], or graph neural network (GNN) models, e.g., GCN [7], to generate structural embeddings of entities in individual KGs. Then, these separated embeddings are projected into a unified embedding space by using the seed entity pairs as connections, so that the entities from different KGs are directly comparable. Finally, to determine the alignment results, the majority of current works [1, 8–10] formalize the alignment process as a ranking problem; that is, for each entity in the source KG, they rank all the entities in the target KG according to some distance metric, and the closest entity is considered as the equivalent target entity.

© The Author(s) 2023
X. Zhao et al., *Entity Alignment*, Big Data Management,
https://doi.org/10.1007/978-981-99-4250-3_8

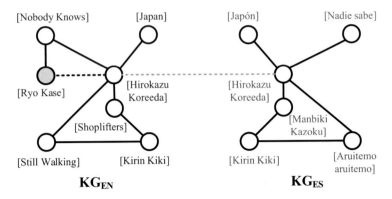

Fig. 8.1 An example of EA

Example In Fig. 8.1 are a partial English KG and a partial Spanish KG concerning the director `Hirokazu Koreeda`, where the dashed lines indicate known alignments (i.e., seeds). The task of EA aims to identify equivalent entity pairs between two KGs, e.g., (`Shoplifters`, `Manbiki Kazoku`).

Nevertheless, we still observe several issues from current EA works:

- **Reliance on labeled data**. Most of the approaches rely on pre-aligned seed entity pairs to connect two KGs and use the unified KG structural embeddings to align entities. These labeled data, however, might not exist in real-life settings. For instance, in the example, the equivalence between `Hirokazu Koreeda` in KG_{EN} and `Hirokazu Koreeda` in KG_{ES} might not be known in advance. In this case, state-of-the-art methods that solely rely on the structural information would fall short, as there are no seeds to connect these individual KGs.

- **Closed-domain setting**. All of current EA solutions work under the closed-domain setting [11]; that is, they assume every entity in the source KG has an equivalent entity in the target KG. Nevertheless, in practical settings, there always exist unmatchable entities. For instance, in the example, for the source entity `Ryo Kase`, there is no equivalent entity in the target KG. Therefore, an ideal EA system should be capable of predicting the unmatchable entities.

In response to these issues, we put forward an unsupervised EA solution UEA that is capable of addressing the unmatchable problem. Specifically, to mitigate the reliance on labeled data, we mine useful features from the KG side information and use them to produce preliminary pseudo-labeled data. These preliminary seeds are forwarded to our devised **progressive learning framework** to generate unified KG structural representations, which are integrated with the side information to provide a more comprehensive view for alignment. This framework also progressively augments the training data and improves the alignment results in a self-training fashion. Besides, to tackle the unmatchable issue, we design an **unmatchable entity prediction** module, which leverages thresholded bidirectional nearest neighbor search (TBNNS) to filter out the unmatchable entities and excludes them from the alignment results. We embed the unmatchable entity prediction module into the progressive learning framework to control the pace of progressive learning by dynamically adjusting the thresholds in TBNNS.

Furthermore, considering that the pseudo-labeled data generated during the progressive learning process might be of different qualities, we introduce the concept of **confidence** to measure the probability of an entity pair of being correct. We further incorporate such confidence scores into KG representation learning with the aim of producing more accurate structural embeddings. Through empirical studies, we demonstrate that the confidence-based framework, CUEA, has a more stable performance than UEA regardless of the quality of input side information and is particularly more useful when the side information is low-grade.

Contribution The main contributions of the chapter can be summarized as follows:

- We identify the deficiencies of existing EA methods, i.e., requiring labeled data and working under the closed-domain setting, and propose an unsupervised EA framework UEA, as well as a confidence-based extension CUEA, that are able to deal with unmatchable entities. This is done by: (1) exploiting the side information of KGs to generate preliminary pseudo-labeled data; and (2) devising an unmatchable entity prediction module that leverages the (confidence-based) thresholded bi-directional nearest neighbor search strategy to produce alignment results, which can effectively exclude unmatchable entities; and (3) offering a progressive learning algorithm to improve the quality of KG embeddings and enhance the alignment performance.
- We empirically evaluate our proposals against state-of-the-art methods, and the comparative results demonstrate their superiority.

Organization In Sect. 8.2, we formally define the task of EA and introduce related work. Section 8.3 elaborates the framework. In Sect. 8.4, we introduce experimental results and conduct detailed analysis. Section 8.5 concludes this chapter.

8.2 Task Definition and Related Work

In this section, we formally define the task of EA and then introduce the related work.

Task Definition The inputs to EA are a source KG G_1 and a target KG G_2. The task of EA is defined as finding the equivalent entities between the KGs, i.e., $\Psi = \{(u, v) | u \in E_1, v \in E_2, u \leftrightarrow v\}$, where E_1 and E_2 refer to the entity sets in G_1 and

G_2, respectively, and $u \leftrightarrow v$ represents that the source entity u and the target entity v are *equivalent*, i.e., u and v refer to the same real-world object.

Most of current EA solutions assume that there exist a set of seed entity pairs $\Psi_s = \{(u_s, v_s) | u_s \in E_1, v_s \in E_2, u_s \leftrightarrow v_s\}$. Nevertheless, in this chapter, we focus on unsupervised EA and do not assume the availability of such labeled data.

Unsupervised Entity Alignment A few methods have investigated the alignment without labeled data. Qu et al. [20] propose an unsupervised approach toward knowledge graph alignment with the adversarial training framework. Nevertheless, the experimental results are extremely poor. He et al. [21] utilize the shared attributes between heterogeneous KGs to generate aligned entity pairs, which are used to detect more equivalent attributes. They perform entity alignment and attribute alignment alternately, leading to more high-quality aligned entity pairs, which are used to train a relation embedding model. Finally, they combine the alignment results generated by attribute and relation triples using a bivariate regression model. The overall procedure of this work might seem similar to our proposed model. However, there are many notable differences; for instance, the KG embeddings in our work are updated progressively, which can lead to more accurate alignment results, and our model can deal with unmatchable entities. We empirically demonstrate the superiority of our model in Sect. 8.4.

We notice that there are some entity resolution (ER) approaches established in a setting similar to EA, represented by PARIS [22]. They adopt collective alignment algorithms such as similarity propagation so as to model the relations among entities. We include them in the experimental study for the comprehensiveness of the chapter.

8.3 Methodology

In this section, we first introduce the outline of our proposal. Then, we elaborate the processing of side information to produce preliminary alignment seeds.

8.3.1 Model Outline

As shown in Fig. 8.2, given two KGs, CUEA first mines useful features from the *side information*. These features are forwarded to the *unmatchable entity prediction* module to generate initial alignment results with confidence scores, which are regarded as pseudo-labeled data. Then, the *progressive learning framework* uses these pseudo seeds, along with the probability scores, to connect two KGs and learn unified entity structural embeddings. It further combines the alignment signals from the side information and *structural information* to provide a more comprehensive view for alignment. Finally, it progressively improves the quality of structural

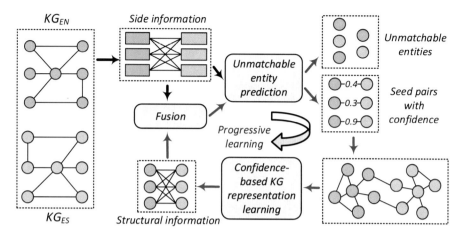

Fig. 8.2 Outline of CUEA. Arrows in blue represent the progressive learning process. By setting the confidence to 1, the UEA model can be restored

embeddings and augments the alignment results by iteratively updating the pseudo-labeled data with results from the previous round, which also leads to increasingly better alignment. Note that by assigning the confidence score of 1 to all entity pairs, CUEA turns into the UEA model.

8.3.2 Side Information

There is abundant side information in KGs, such as the attributes, descriptions, and classes. In this chapter, we use a particular form of the attributes—the entity name, as it exists in the majority of KGs. To make the most of the entity name information, inspired by Zeng et al. [5], we exploit it from the semantic level and string level and generate the textual distance matrix between entities in two KGs.

More specifically, we use the averaged word embeddings to represent the semantic meanings of entity names. Given the semantic embeddings of a source and a target entity, we obtain the semantic distance score by subtracting their cosine similarity score from 1. We denote the semantic distance matrix between the entities in two KGs as $\mathbf{M^n}$, where rows represent source entities, columns denote target entities, and each element in the matrix denotes the distance score between a pair of source and target entities. As for the string-level feature, we adopt the Levenshtein distance [23] to measure the difference between two sequences. We denote the string distance matrix as $\mathbf{M^l}$.

To obtain a more comprehensive view for alignment, we combine these two distance matrices and generate the textual distance matrix as $\mathbf{M^t} = \alpha\mathbf{M^n} + (1 - \alpha)\mathbf{M^l}$, where α is a hyper-parameter that balances the weights. Then, we forward the textual distance matrix $\mathbf{M^t}$ into the unmatchable entity module to produce alignment results, which are considered as the pseudo-labeled data for training KG structural embeddings. The details are introduced in the next subsection.

Remark The goal of this step is to exploit available side information to generate useful features for alignment. Other types of side information, e.g., attributes and entity descriptions, can also be leveraged. Besides, more advanced textual encoders, such as misspelling oblivious word embeddings [24] and convolutional embedding for edit distance [25], can be utilized. We will investigate them in the future.

8.3.3 Unmatchable Entity Prediction

State-of-the-art EA solutions generate for each source entity a corresponding target entity and fail to consider the potential unmatchable issue. Nevertheless, as mentioned in [12], in real-life settings, KGs contain entities that other KGs do not contain. For instance, when aligning YAGO 4 and IMDB, only 1% of entities in YAGO 4 are related to movies, while the other 99% of entities in YAGO 4 necessarily have no match in IMDB. These unmatchable entities would increase the difficulty of EA. Therefore, in this chapter, we devise an unmatchable entity prediction module to predict the unmatchable entities and filter them out from the alignment results.

8.3.3.1 Thresholded Bidirectional Nearest Neighbor Search

More specifically, we put forward a novel strategy, i.e., thresholded bidirectional nearest neighbor search (TBNNS), to generate the alignment results, and the resulting unaligned entities are predicted to be unmatchable. As can be observed from Algorithm 1, given a source entity u and a target entity v, if u and v are the nearest neighbor of each other, and the distance between them is below a given threshold θ, we consider (u, v) as an aligned entity pair. Note that $\mathbf{M}(u, v)$ represents the element in the u-th row and v-th column of the distance matrix $\mathbf{M}$.

Algorithm 1: TBNNS in the unmatchable entity prediction module

---

Input   : G_1 and G_2: the two KGs to be aligned; E_1 and E_2: the entity sets in G_1 and G_2;
             θ: a given threshold; $\mathbf{M}$: a distance matrix.

Output : S: Alignment results.

1 **foreach** $u \in E_1$ **do**

2   $\quad v \leftarrow \arg\min_{\hat{v} \in E_2} \mathbf{M}(u, \hat{v})$;

3   $\quad$ **if** $\arg\min_{\hat{u} \in E_1} \mathbf{M}(v, \hat{u}) = u$ **and** $\mathbf{M}(u, v) < \theta$ **then**

4   $\quad\quad S \leftarrow S + \{(u, v)\}$

5 **return** S.

The TBNNS strategy exerts strong constraints on alignment, since it requires that the matched entities should both prefer each other the most, and the distance between their embeddings should be below a certain value. Therefore, it can effectively predict unmatchable entities and prevent them from being aligned. Notably, the threshold θ plays a significant role in this strategy. A larger threshold would lead to more matches, whereas it would also increase the risk of including erroneous matches or unmatchable entities. In contrast, a small threshold would only lead to a few aligned entity pairs, and almost all of them would be correct. This is further discussed and verified in Sect. 8.4.4. Therefore, our progressive learning framework dynamically adjusts the threshold value to produce more accurate alignment results (to be discussed in the next subsection).

8.3.3.2 Confidence-Based TBNNS

Considering that the aligned entity pairs generated by TBNNS are of different qualities (i.e., some are true, while some are not), we further put forward confidence-based TBNNS, C-TBNNS, to measure the confidence of an entity pair (of being true). Specifically, we define the confidence score Θ of an entity pair (u, v) as:

$$\Theta(u, v) = \mathbf{M}(u, v') - \mathbf{M}(u, v) + \mathbf{M}(v, u') - \mathbf{M}(v, u), \qquad (8.1)$$

where $\Delta_1 = \mathbf{M}(u, v') - \mathbf{M}(u, v)$ denotes the gap between the distance scores of the top two closest entities (i.e., v and v') to entity u, while $\Delta_2 = \mathbf{M}(v, u') - \mathbf{M}(v, u)$ denotes the gap between the distance scores of the top two closest entities (i.e., u and u') to entity v. This is based on the intuition that, for an entity pair (u, v), if the distance between them is the smallest from both sides and there are larger margins between the distances of the top two candidates, it would be more confident to consider them as a correct entity pair. We further restrict the confidence scores to a certain range:

$$\Theta(\mathcal{S}) = (1 - \lambda) \frac{\Theta(\mathcal{S}) - \min\{\Theta(\mathcal{S})\}}{\max\{\Theta(\mathcal{S})\} - \min\{\Theta(\mathcal{S})\}} + \lambda \qquad (8.2)$$

where $\Theta(\mathcal{S})$ represents the confidence scores of the entity pairs in $\mathcal{S}$. The core of Eq. (8.2) is the min-max normalization, which converts the confidence scores to [0, 1]. We add a hyper-parameter $\lambda \in [0, 1]$ to further restrict the range of the confidence scores to $[\lambda, 1]$. As thus, by setting λ to 1, all entity pairs would have the same confidence score of 1, and C-TBNNS can be restored to TBNNS. Hence, C-TBNNS can be regarded as a general case of TBNNS, which introduces the concept of confidence (probability) into the alignment result generation process.

8.3.4   The Progressive Learning Framework

To exploit the rich structural patterns in KGs that could provide useful signals for alignment, we design a progressive learning framework to combine structural and textual features for alignment and improve the quality of both structural embeddings and alignment results in a self-training fashion.

8.3.4.1   Knowledge Graph Representation Learning

As mentioned above, we forward the textual distance matrix $\mathbf{M^t}$ generated by using the side information to the unmatchable entity prediction module to produce the preliminary alignment results, which are considered as pseudo-labeled data for learning unified KG embeddings. Concretely, following [18], we adopt GCN[1] to capture the neighboring information of entities. We leave out the implementation details since this is not the focus of this paper, which can be found in [18].

Alignment Objective Since the representations of source and target KGs are learned individually, they need to be projected into a unified embedding space, where the entities across KGs could be compared directly. To this end, we use the semi-supervised loss function to enforce the distance between the embeddings of the entities in the labeled entity pairs to be small and meanwhile the negative samples (i.e., nonequivalent entity pairs) to be large. Formally:

$$\mathcal{L} = \sum_{(u,v)\in\mathcal{S}} \sum_{(u',v')\in\mathcal{S}'_{(u,v)}} [d(\mathbf{u}, \mathbf{v}) + \gamma - d(\mathbf{u}', \mathbf{v}')]_+, \qquad (8.3)$$

where $[\cdot]_+ = \max\{0, \cdot\}$, (u, v) is a labeled entity pair from the training data and $\mathcal{S}'_{(u,v)}$ represents the set of negative entity pairs obtained by corrupting (u, v) using nearest neighbor sampling [1]. $\mathbf{u}$ and $\mathbf{v}$ represent the embeddings of source and target entities learned by GCN, respectively. $d(\cdot, \cdot)$ is the distance function that measures the distance between two embeddings. γ is a hyper-parameter separating positive samples from negative ones.

Confidence-Based Objective Considering that the pseudo-labeled entity pairs have different confidences of being true, we incorporate such probabilities into the

---

[1] More advanced structural learning models, such as recurrent skipping networks [13], could also be used here. We will explore these alternative options in the future.

alignment objective to learn more accurate structural embeddings:

$$\mathcal{L}_c = \sum_{(u,v)\in\mathcal{S}} \sum_{(u',v')\in\mathcal{S}'_{(u,v)}} \Theta(u,v) * [d(\mathbf{u},\mathbf{v}) + \gamma - d(\mathbf{u}',\mathbf{v}')]_+, \qquad (8.4)$$

where $\Theta(u,v)$ is the confidence score attached to each entity pair. As thus, the more confident entity pairs would play a more important role during the training process, while the less confident pseudo entity pairs would have a smaller effect on the training, such that the impact from the false positives could be mitigated.

Feature Fusion Given the learned structural embedding matrix $\mathbf{Z}$, we calculate the structural distance score between a source and a target entity by subtracting the cosine similarity score between their embeddings from 1. We denote the resultant structural distance matrix as $\mathbf{M}^s$. Then, we combine the textual and structural information to generate more accurate signals for alignment: $\mathbf{M} = \beta\mathbf{M}^t + (1-\beta)\mathbf{M}^s$, where β is a hyper-parameter that balances the weights. The fused distance matrix $\mathbf{M}$ can be used to generate more accurate matches.

8.3.4.2 The Progressive Learning Algorithm

The amount of training data has an impact on the quality of the unified KG embeddings, which in turn affects the alignment performance [3, 26]. As thus, we devise an algorithm (Algorithm 2) to progressively augment the pseudo training data, so as to improve the quality of KG embeddings and enhance the alignment performance. The algorithm starts with learning unified structural embeddings and generating the fused distance matrix $\mathbf{M}$ by using the preliminary pseudo-labeled data $\mathcal{S}_0$ (Lines 1–2). Then, the fused distance matrix is used to produce the new alignment results $\Delta\mathcal{S}$ using C-TBNNS (line 4). These newly generated entity pairs $\Delta\mathcal{S}$ are added to the alignment results, which are used for generating the fused distance matrix in the next round (Lines 6–7). The entities in $\mathcal{S}$ are removed from the entity sets (Lines 9–10). In order to progressively improve the quality of KG embeddings and detect more alignment results, we perform the aforementioned process recursively until the number of newly generated entity pairs is below a given threshold μ. Finally, we consider the entity pairs in $\mathcal{S}$ as the final alignment results Ψ.

Notably, in the learning process, once a pair of entities is considered as a match, the entities will be removed from the entity sets (Lines 5–6 and Lines 12–13). This could gradually reduce the alignment search space and lower the difficulty for aligning the rest of the entities. Obviously, this strategy suffers from the error propagation issue, which, however, could be effectively mitigated by the progressive learning process that dynamically adjusts the threshold. We will verify the effectiveness of this setting in Sect. 8.4.3.

Algorithm 2: Progressive learning

Input   : $\mathcal{G}_1$ and $\mathcal{G}_2$: KGs to be aligned; $\mathcal{E}_1$ and $\mathcal{E}_2$: the entity sets; $\mathbf{M^t}$: textual distance
matrix; S_0: preliminary labeled data; θ_0: the initial threshold.
Output : Ψ: Alignment results.
1 $S \leftarrow S_0$;
2 Use S to learn structural embeddings and generate $\mathbf{M}$;
3 $\theta \leftarrow \theta_0$;
4 $\Delta S, \mathcal{U} \leftarrow$ C-TBNNS $(\mathcal{G}_1, \mathcal{G}_2, \mathcal{E}_1, \mathcal{E}_2, \theta, \mathbf{M})$;
5 **while** $|\Delta S| \geq \mu$ **do**
6 $\quad$ $S \leftarrow S + \Delta S$;
7 $\quad$ Use S to learn structural embeddings and generate $\mathbf{M}$;
8 $\quad$ $\theta \leftarrow \theta + \eta$;
9 $\quad$ $\mathcal{E}_1 \leftarrow \{e | e \in \mathcal{E}_1, e \notin S\}$;
10 $\quad$ $\mathcal{E}_2 \leftarrow \{e | e \in \mathcal{E}_2, e \notin S\}$;
11 $\quad$ $\Delta S, \mathcal{U} \leftarrow$ C-TBNNS $(\mathcal{G}_1, \mathcal{G}_2, \mathcal{E}_1, \mathcal{E}_2, \theta, \mathbf{M})$;
12 $\Psi \leftarrow S$;
13 **return** Ψ.

8.3.4.3   Dynamic Threshold Adjustment

It can be observed from Algorithm 2 that the matches generated by the unmatchable entity prediction module are part of not only the eventual alignment results but also the pseudo training data for learning subsequent structural embeddings. Therefore, to enhance the overall alignment performance, the alignment results generated in each round should, ideally, have both large *quantity* and high *quality*. Unfortunately, these two goals cannot be achieved at the same time. This is because, as stated in Sect. 8.3.3, a larger threshold in TBNNS can generate more alignment results (large quantity), whereas some of them might be erroneous (low quality). These wrongly aligned entity pairs can cause the error propagation problem and result in more erroneous matches in the following rounds. In contrast, a smaller threshold leads to fewer alignment results (small quantity), while almost all of them are correct (high quality).

To address this issue, we aim to balance between the quantity and the quality of the matches generated in each round. An intuitive idea is to set the threshold to a moderate value. However, this fails to take into account the characteristics of the progressive learning process. That is, in the beginning, the quality of the matches should be prioritized, as these alignment results will have a long-term impact on the subsequent rounds. In comparison, in the later stages where most of the entities have been aligned, the quantity is more important, as we need to include more possible matches that might not have a small distance score. In this connection, we set the initial threshold θ_0 to a very small value so as to reduce potential errors. Then, in the following rounds, we gradually increase the threshold by η, so that more possible matches could be detected. We will empirically validate the superiority of this strategy over the fixed weight in Sect. 8.4.3.

Noteworthily, our proposed confidence-based framework CUEA can further help mitigate the low-quality issue, as we calculate and assign a confidence score to each entity pair, where the wrongly aligned entity pairs would presumably have lower confidence scores and thus exert smaller influence on the subsequent alignment process.

Remark As mentioned in the related work, there are some existing EA approaches that exploit the iterative learning (bootstrapping) strategy to improve EA performance. Particularly, BootEA calculates for each source entity the alignment likelihood to every target entity and includes those with likelihood above a given threshold in a maximum likelihood matching process under the 1-to-1 mapping constraint, producing a solution containing confident EA pairs [15]. This strategy is also adopted by [8, 16]. Zhu et al. use a threshold to select the entity pairs with very close distances as the pseudo-labeled data [14]. DAT employs a bidirectional margin-based constraint to select the confident EA pairs as labels [17]. Our progressive learning strategy differs from these existing solutions in three aspects: (1) we exclude the entities in the confident EA pairs from the test sets; (2) we use the dynamic threshold adjustment strategy to control the pace of learning process; (3) our strategy can deal with unmatchable entities; and (4) we attach a confidence score to each selected entity pair, which can mitigate the negative influence of the false positives on the KG representation learning process as well as the alignment results. The superiority of our strategy is validated in Sect. 8.4.3.

8.4 Experiment

This section reports the experimental results with in-depth analysis. The source code is available at https://github.com/DexterZeng/UEA.

8.4.1 Experimental Settings

Datasets Following existing works, we adopt the DBP15K dataset [3] for evaluation. This dataset consists of three multilingual KG pairs extracted from DBpedia. Each KG pair contains 15,000 inter-language links as gold standards. The statistics can be found in Table 8.1. We note that state-of-the-art studies merely consider the labeled entities and divide them into training and testing sets. Nevertheless, as can be observed from Table 8.1, there exist unlabeled entities, e.g., 4,388 and 4,572 entities in the Chinese and English KG of DBP15K$_{ZH-EN}$, respectively. In this connection, we adapt the dataset by including the unmatchable entities. Specifically, for each KG pair, we keep 30% of the labeled entity pairs as the training set (for training the supervised or semi-supervised methods). Then, to construct the test set, we include the rest of the entities in the first KG and the rest of the labeled entities in the second

Table 8.1 The statistics of the evaluation benchmarks

Dataset	KG pairs	#Triples	#Entities	#Labeled Ents	#Relations	#Test set
$DBP15K_{ZH-EN}$	DBpedia(Chinese)	70,414	19,388	15,000	1,701	14,888
	DBpedia(English)	95,142	19,572	15,000	1,323	10,500
$DBP15K_{JA-EN}$	DBpedia(Japanese)	77,214	19,814	15,000	1,299	15,314
	DBpedia(English)	93,484	19,780	15,000	1,153	10,500
$DBP15K_{FR-EN}$	DBpedia(French)	105,998	19,661	15,000	903	15,161
	DBpedia(English)	115,722	19,993	15,000	1,208	10,500

KG, so that the unlabeled entities in the first KG become unmatchable. The statistics of the test sets can be found in the *test set* column in Table 8.1.

Parameter Settings For the *side information* module, we utilize the fastText embeddings [27] as word embeddings. To deal with cross-lingual KG pairs, following [19], we use Google Translate to translate the entity names from one language to another, i.e., translating Chinese, Japanese, and French to English. α is set to 0.5. For the *structural information learning*, we set β to 0.5. Following [18], we set γ in the alignment objectives to 3 and adopt Manhattan distance as $d(\cdot, \cdot)$. Regarding C-TBNNS, we set λ to 0.4. For *progressive learning*, we set the initial threshold θ_0 to 0.05, the incremental parameter η to 0.1, and the termination threshold γ to 30. Note that if the threshold θ is over 0.45, we reset it to 0.45. These hyper-parameters are default values since there is no extra validation set for hyper-parameter tuning.

Evaluation Metrics We use *precision* (P), *recall* (R), and *F1 score* as evaluation metrics. The *precision* is computed as the number of correct matches divided by the number of matches found by a method. The *recall* is computed as the number of correct matches found by a method divided by the number of gold matches. The *F1 score* is the harmonic mean between *precision* and *recall*. The bold figures in the tables represent the best results.

Competitors We select the most performant state-of-the-art solutions for comparison. Within the group that solely utilizes structural information, we compare with BootEA [15], TransEdge [8], MRAEA [26], and SSP [28]. Among the methods incorporating other sources of information, we compare with GCN-Align [18], HMAN [9], HGCN [4], RE-GCN [29], DAT [17], and RREA [30]. We also include the unsupervised approaches, i.e., IMUSE [21] and PARIS [22]. To make a fair comparison, we only use entity name labels as the side information.

Table 8.2 Alignment results

	DBP15K$_{\text{ZH-EN}}$			DBP15K$_{\text{JA-EN}}$			DBP15K$_{\text{FR-EN}}$			SRPRS$_{\text{EN-FR}}$			SRPRS$_{\text{EN-DE}}$		
	P	R	F1	P	R	F1	P	R	F1	P	R	F1	P	R	F1
BootEA	0.444	0.629	0.520	0.426	0.622	0.506	0.452	0.653	0.534	0.365	0.365	0.365	0.503	0.503	0.503
TransEdge	0.518	0.735	0.608	0.493	0.719	0.585	0.492	0.710	0.581	0.400	0.400	0.400	0.556	0.556	0.556
MRAEA	0.534	0.757	0.626	0.520	0.758	0.617	0.540	0.780	0.638	0.403	0.403	0.403	0.543	0.543	0.543
SSP	0.521	0.739	0.611	0.494	0.721	0.587	0.512	0.739	0.605	0.372	0.372	0.372	0.521	0.521	0.521
RREA	0.565	0.801	0.663	0.550	0.802	0.652	0.573	0.827	0.677	0.468	0.468	0.468	0.601	0.601	0.601
GCN-Align	0.291	0.413	0.342	0.274	0.399	0.325	0.258	0.373	0.305	0.296	0.758	0.426	0.428	0.428	0.428
HMAN	0.614	0.871	0.720	0.641	0.935	0.761	0.674	**0.973**	0.796	0.400	0.400	0.400	0.528	0.528	0.528
HGCN	0.508	0.720	0.596	0.525	0.766	0.623	0.618	0.892	0.730	0.670	0.670	0.670	0.763	0.763	0.763
RE-GCN[a]	0.518	0.735	0.608	0.548	0.799	0.650	0.646	0.933	0.764	–	–	–	–	–	–
DAT	0.556	0.788	0.652	0.573	0.835	0.679	0.639	0.922	0.755	0.758	0.758	0.758	0.876	0.876	0.876
IMUSE	0.608	0.862	0.713	0.625	0.911	0.741	0.618	0.892	0.730	0.905	0.905	0.905	0.916	0.916	0.916
PARIS	**0.976**	0.777	0.865	**0.981**	0.785	0.872	**0.972**	0.793	0.873	0.990	0.870	0.926	0.990	0.930	0.959
UEA	0.913	**0.902**	**0.907**	0.940	0.932	0.936	0.953	0.950	**0.951**	0.987	0.969	0.978	**0.988**	**0.976**	**0.982**
CUEA	0.912	0.901	0.906	0.943	**0.935**	**0.939**	0.953	0.949	**0.951**	**0.988**	**0.970**	**0.979**	0.988	0.975	0.981

We omit the results of RE-GCN on SRPRS$_{\text{EN-FR}}$ and SRPRS$_{\text{EN-DE}}$ since they are not provided in the original paper, and our implementation cannot reproduce the reported performance

On SRPRS$_{\text{EN-FR}}$ and SRPRS$_{\text{EN-DE}}$, all of the entities are matchable, and the number of entities in a KG. Besides, for most methods, they generate matches for all the entities in a KG. Therefore, the number of matches produced by these methods is equal to the number of gold matches, and the values of precision, recall, and F1 score are equal

8.4.2 Results

Table 8.2 reports the alignment results, which shows that state-of-the-art supervised
or semi-supervised methods have rather low precision values. This is because these
approaches cannot predict the unmatchable source entities and generate a target
entity for each source entity (including the unmatchable ones). Particularly, methods
incorporating additional information attain relatively better performance than the
methods in the first group, demonstrating the benefit of leveraging such additional
information.

Regarding the unsupervised methods, although IMUSE cannot deal with the
unmatchable entities and achieves a low precision score, it outperforms most of
the supervised or semi-supervised methods in terms of recall and F1 score. This
indicates that, for the EA task, the KG side information is useful for mitigating
the reliance on labeled data. In contrast to the abovementioned methods, PARIS
attains very high precision, since it only generates matches that it believes to
be highly possible, which can effectively filter out the unmatchable entities. It
also achieves the second best F1 score among all approaches, showcasing its
effectiveness when the unmatchable entities are involved. Our proposals, UEA
and CUEA, attain the best balance between precision and recall and obtain the
best F1 scores, outperforming the second best by a large margin, validating their
effectiveness. Notably, although our proposed models do not require labeled data,
they achieve even better performance than the most performant supervised methods
HMAN and DAT.

Furthermore, it can be seen that, by integrating the notion of confidence into UEA,
CUEA achieves comparable results to UEA. At first sight, it seems that assigning
confidence scores to entity pairs does not have a large influence on the representation
learning and the alignment results, which, however, could be ascribed to the fact
that the side information is too effective on these datasets (solely using the string
information can achieve an F1 score of 0.814, to be shown in Table 8.4), and hence
rendering the structural information (largely affected by the confidence scores) less
contributive to the overall results. Next, we will show that the confidence-based
framework would be much more useful on datasets with side information in low
quality.

8.4.2.1 Results Using Low-Quality Side Information

We compare the unsupervised approaches under a practical scenario where the side
information is in low quality. Specifically, we assume that the pre-trained word
embeddings as well as the machine translation tools are not available. Under this
circumstance, to use the entity name information, a viable solution is to compare the
name strings directly. However, the direct string comparison would be ineffective
for cross-lingual datasets such as DBP15K$_{ZH-EN}$ and DBP15K$_{JA-EN}$, where the
languages in the source and target KGs are disparate. Hence, we aim to examine

Table 8.3 Alignment results given low-grade side information

	DBP15K$_{ZH-EN}$			DBP15K$_{JA-EN}$		
	P	R	F1	P	R	F1
IMUSE	0.056	0.080	0.066	0.053	0.077	0.063
PARIS	**0.921**	0.066	0.123	**0.911**	0.060	0.113
UEA	0.654	0.088	0.155	0.617	0.084	0.148
CUEA	0.682	**0.093**	**0.164**	0.690	**0.090**	**0.159**

the effectiveness of these unsupervised approaches when the side information is in low quality and cannot provide many useful signals for alignment.

We report the results on DBP15K$_{ZH-EN}$ and DBP15K$_{JA-EN}$ in Table 8.3, where the direct comparison between entity name strings serves as the side information. It can be observed that the F1 scores of all methods are very low (compared with those in Table 8.2), revealing that the quality of side information does affect the overall alignment results. Besides, given the low-quality side information, our proposed models UEA and CUEA still outperform the baselines IMUSE and PARIS in terms of the F1 score, demonstrating the effectiveness of the progressive learning framework and the unmatchable entity prediction module. Moreover, it is notable that CUEA achieves better results than UEA in terms of all metrics. This could be attributed to the confidence-based alignment result generation process, which could enable the entity pairs of higher confidence (higher probability of being correct, presumably) to have a larger impact on the representation learning and alignment process.

Table 8.4 Ablation results

	DBP15K$_{ZH-EN}$			DBP15K$_{JA-EN}$			DBP15K$_{FR-EN}$		
	P	R	F1	P	R	F1	P	R	F1
UEA	0.913	0.902	**0.907**	0.940	0.932	**0.936**	0.953	0.950	**0.951**
w/o Unm	0.553	0.784	0.648	0.578	0.843	0.686	0.603	0.871	0.713
w/o Prg	0.942	0.674	0.786	0.966	0.764	0.853	0.972	0.804	0.880
w/o Adj	0.889	0.873	0.881	0.927	0.915	0.921	0.941	0.936	0.939
w/o Excl	**0.974**	0.799	0.878	0.982	0.862	0.918	0.985	0.887	0.933
MWGM	0.930	0.789	0.853	0.954	0.858	0.903	0.959	0.909	0.934
TH	0.743	**0.914**	0.820	0.795	**0.942**	0.862	0.807	**0.953**	0.874
DAT-I	**0.974**	0.805	0.881	**0.985**	0.866	0.922	**0.988**	0.875	0.928
UEA-M^l	0.908	0.902	0.905	0.926	0.924	0.925	0.937	0.931	0.934
M^l	0.935	0.721	0.814	0.960	0.803	0.875	0.948	0.750	0.838
UEA-M^n	0.758	0.727	0.742	0.840	0.807	0.823	0.906	0.899	0.903
M^n	0.891	0.497	0.638	0.918	0.562	0.697	0.959	0.752	0.843

8.4.3 Ablation Study

In this subsection, we examine the usefulness of proposed modules by conducting the ablation study. More specifically, in Table 8.4, we report the results of UEA w/o Unm, which excludes the unmatchable entity prediction module, and UEA w/o Prg, which excludes the progressive learning process. It shows that removing the unmatchable entity prediction module (UEA w/o Unm) brings down the performance on all metrics and datasets, validating its effectiveness of detecting the unmatchable entities and enhancing the overall alignment performance. Besides, without the progressive learning (UEA w/o Prg), the precision increases, while the recall and F1 score values drop significantly. This shows that the progressive learning framework can discover more correct aligned entity pairs and is crucial to the alignment progress.

To provide insights into the progressive learning framework, we report the results of UEA w/o Adj, which does not adjust the threshold, and UEA w/o Excl, which does not exclude the entities in the alignment results from the entity sets during the progressive learning. Table 8.4 shows that setting the threshold to a fixed value (UEA w/o Adj) leads to worse F1 results, verifying that the progressive learning process depends on the choice of the threshold and the quality of the alignment results. We will further discuss the setting of the threshold in the next subsection. Besides, the performance also decreases if we do not exclude the matched entities from the entity sets (UEA w/o Excl), validating that this strategy indeed can reduce the difficulty of aligning entities.

Moreover, we replace our progressive learning framework with other state-of-the-art iterative learning strategies (i.e., MWGM [15], TH [14], and DAT-I [17]) and report the results in Table 8.4. It shows that using our progressive learning framework (UEA) can attain the best F1 score, verifying its superiority.

8.4.4 Quantitative Analysis

In this subsection, we perform quantitative analysis of the modules in UEA and CUEA.

The Threshold θ in TBNNS We discuss the setting of θ to reveal the trade-off between the risk and gain from generating the alignment results in the progressive learning. Identifying a match leads to the integration of additional structural information, which benefits the subsequent learning. However, for the same reason, the identification of a false positive, i.e., an incorrect match, potentially leads to mistakenly modifying the connections between KGs, with the risk of amplifying the error in successive rounds. As shown in Fig. 8.3, a smaller θ (e.g., 0.05) brings low risk and low gain; that is, it merely generates a small number of matches, among which almost all are correct. In contrast, a higher θ (e.g., 0.45) increases the risk and brings relatively higher gain; that is, it results in much more aligned entity

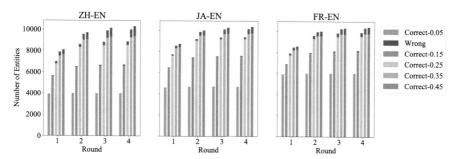

Fig. 8.3 Alignment results given different threshold values. Correct-θ refers to the number of correct matches generated by the progressive learning framework at each round given the threshold value θ. Wrong refers to the number of erroneous matches generated in each round

pairs, while a certain portion of them are erroneous. Additionally, using a higher threshold leads to increasingly more alignment results, while for a lower threshold, the progressive learning process barely increases the number of matches. This is in consistency with our theoretical analysis in Sect. 8.3.3.

Unmatchable Entity Prediction Zhao et al. [12] propose an intuitive strategy (U-TH) to predict the unmatchable entities. They set an NIL threshold, and if the distance value between a source entity and its closest target entity is above this threshold, they consider the source entity to be unmatchable. We compare our unmatchable entity prediction strategy with it in terms of the percentage of unmatchable entities that are included in the final alignment results and the F1 score. On DBP15K$_{\text{ZH-EN}}$, replacing our unmatchable entity prediction strategy with U-TH attains the F1 score at 0.837, which is 8.4% lower than that of UEA. Besides, in the alignment results generated by using U-TH, 18.9% are unmatchable entities, while this figure for UEA is merely 3.9%. This demonstrates the superiority of our unmatchable entity prediction strategy.

Influence of Parameters As mentioned in Sect. 8.4.1, we set α and β to 0.5 since there are no training/validation data. Here, we aim to prove that different values of the parameters do not have a large influence on the final results. More specifically, we keep α at 0.5 and choose β from [0.3, 0.4, 0.5, 0.6, 0.7]; then we keep β at 0.5 and choose α from [0.3, 0.4, 0.5, 0.6, 0.7]. It can be observed from Fig. 8.4 that, although smaller α and β lead to better results, the performance does not change significantly.

The Hyper-Parameter λ in CUEA We then analyze the influence of λ in Eq. (8.2), which determines the range of the confidence scores, on the final alignment results. To highlight its influence on the structural representation learning, we follow the settings in Sect. 8.4.2.1 and report the results in Table 8.5.

It reads from Table 8.5 that the alignment performance is relatively stable when λ is not too large. Nevertheless, when setting λ to a large value (e.g., 1, to restore

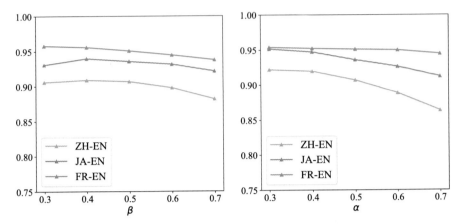

Fig. 8.4 The F1 scores by setting α and β to different values

Table 8.5 The influence of λ on the alignment results

	DBP15K$_{\text{ZH-EN}}$			DBP15K$_{\text{JA-EN}}$		
	P	R	F1	P	R	F1
CUEA	0.682	**0.093**	**0.164**	**0.690**	**0.090**	**0.159**
$\lambda = 0.0$	0.685	**0.093**	**0.164**	0.680	**0.090**	**0.159**
$\lambda = 0.1$	0.687	0.092	0.162	0.683	**0.090**	**0.159**
$\lambda = 0.2$	**0.689**	**0.093**	0.163	0.680	**0.090**	**0.159**
$\lambda = 0.3$	0.678	0.092	0.162	0.679	0.089	0.158
$\lambda = 0.5$	0.661	0.091	0.160	0.672	**0.090**	**0.159**
$\lambda = 0.6$	0.670	0.090	0.159	0.648	0.088	0.155
$\lambda = 0.7$	0.666	0.089	0.158	0.631	0.085	0.150
$\lambda = 0.8$	0.647	0.088	0.156	0.640	0.088	0.155
$\lambda = 0.9$	0.649	0.088	0.155	0.586	0.081	0.142
$\lambda = 1.0$	0.654	0.088	0.155	0.617	0.084	0.148

UEA), the results drop sharply. This reveals that assigning probability scores to the entity pairs according to their confidence of being true can facilitate the alignment. Besides, generally speaking, CUEA is robust to the perturbation of λ (as long as it is not too large).

Influence of Input Side Information We adopt different side information as input to examine the performance of UEA. More specifically, we report the results of UEA-$\mathbf{M^l}$, which merely uses the string-level feature of entity names as input, UEA-$\mathbf{M^n}$, which only uses the semantic embeddings of entity names as input. We also provide the results of $\mathbf{M^l}$ and $\mathbf{M^n}$, which use the string-level and semantic information to directly generate alignment results (without progressive learning), respectively.

As shown in Table 8.3, the performance of solely using the input side information is not very promising ($\mathbf{M^l}$ and $\mathbf{M^n}$). Nevertheless, by forwarding the side information into our model, the results of UEA-$\mathbf{M^l}$ and UEA-$\mathbf{M^n}$ become much better. This unveils that UEA can work with different types of side information and consistently improve the alignment results. Additionally, by comparing UEA-M^l with UEA-M^n, it is evident that the input side information does affect the final results, and the quality of the side information is of significance to the overall alignment performance.

Pseudo-Labeled Data We further examine the usefulness of the preliminary alignment results generated by the side information, i.e., the pseudo-labeled data. Concretely, we replace the training data in HGCN with these pseudo-labeled data, resulting in HGCN-U, and then compare its alignment results with the original performance. Regarding the F1 score, HGCN-U is 4% lower than HGCN on DBP15K$_{\text{ZH-EN}}$, 2.9% lower on DBP15K$_{\text{JA-EN}}$, and 2.8% lower on DBP15K$_{\text{FR-EN}}$. The minor difference validates the effectiveness of the pseudo-labeled data generated by the side information. It also demonstrates that this strategy can be applied to other supervised or semi-supervised frameworks to reduce their reliance on labeled data.

8.5 Conclusion

In this chapter, we propose an unsupervised EA solution that is capable of dealing with unmatchable entities. We first exploit the side information of KGs to generate preliminary alignment results, which are considered as pseudo-labeled data and forwarded to the progressive learning framework to produce better KG embeddings and alignment results in a self-training fashion. We also devise an unmatchable entity prediction module to detect the unmatchable entities. The experimental results validate the usefulness of our proposed model and its superiority over state-of-the-art approaches.

References

1. Chengjiang Li, Yixin Cao, Lei Hou, Jiaxin Shi, Juanzi Li, and Tat-Seng Chua. Semi-supervised entity alignment via joint knowledge embedding model and cross-graph model. In *EMNLP*, pages 2723–2732, 2019.
2. Muhao Chen, Yingtao Tian, Mohan Yang, and Carlo Zaniolo. Multilingual knowledge graph embeddings for cross-lingual knowledge alignment. In *IJCAI*, pages 1511–1517, 2017.
3. Zequn Sun, Wei Hu, and Chengkai Li. Cross-lingual entity alignment via joint attribute-preserving embedding. In *ISWC*, pages 628–644, 2017.
4. Yuting Wu, Xiao Liu, Yansong Feng, Zheng Wang, and Dongyan Zhao. Jointly learning entity and relation representations for entity alignment. In *EMNLP*, pages 240–249, 2019.
5. Weixin Zeng, Xiang Zhao, Jiuyang Tang, and Xuemin Lin. Collective entity alignment via adaptive features. In *ICDE*, pages 1870–1873, 2020.
6. Antoine Bordes, Nicolas Usunier, Alberto García-Durán, Jason Weston, and Oksana Yakhnenko. Translating embeddings for modeling multi-relational data. In *NIPS*, pages 2787–2795, 2013.
7. Thomas N. Kipf and Max Welling. Semi-supervised classification with graph convolutional networks. *CoRR*, abs/1609.02907, 2016.

8. Zequn Sun, JiaCheng Huang, Wei Hu, Muhao Chen, Lingbing Guo, and Yuzhong Qu. Transedge: Translating relation-contextualized embeddings for knowledge graphs. In *ISWC*, pages 612–629, 2019.
9. Hsiu-Wei Yang, Yanyan Zou, Peng Shi, Wei Lu, Jimmy Lin, and Xu Sun. Aligning cross-lingual entities with multi-aspect information. In *EMNLP*, pages 4430–4440, 2019.
10. Yixin Cao, Zhiyuan Liu, Chengjiang Li, Zhiyuan Liu, Juanzi Li, and Tat-Seng Chua. Multi-channel graph neural network for entity alignment. In *ACL*, pages 1452–1461, 2019.
11. Sven Hertling and Heiko Paulheim. The knowledge graph track at OAEI - gold standards, baselines, and the golden hammer bias. In *ESWC*, volume 12123, pages 343–359, 2020.
12. Xiang Zhao, Weixin Zeng, Jiuyang Tang, Wei Wang, and Fabian Suchanek. An experimental study of state-of-the-art entity alignment approaches. *IEEE Transactions on Knowledge and Data Engineering*, pages 1–1, 2020.
13. Lingbing Guo, Zequn Sun, and Wei Hu. Learning to exploit long-term relational dependencies in knowledge graphs. In *ICML*, pages 2505–2514, 2019.
14. Hao Zhu, Ruobing Xie, Zhiyuan Liu, and Maosong Sun. Iterative entity alignment via joint knowledge embeddings. In *IJCAI*, pages 4258–4264, 2017.
15. Zequn Sun, Wei Hu, Qingheng Zhang, and Yuzhong Qu. Bootstrapping entity alignment with knowledge graph embedding. In *IJCAI*, pages 4396–4402, 2018.
16. Qiannan Zhu, Xiaofei Zhou, Jia Wu, Jianlong Tan, and Li Guo. Neighborhood-aware attentional representation for multilingual knowledge graphs. In *IJCAI*, pages 1943–1949, 2019.
17. Weixin Zeng, Xiang Zhao, Wei Wang, Jiuyang Tang, and Zhen Tan. Degree-aware alignment for entities in tail. In *SIGIR*, pages 811–820, 2020.
18. Zhichun Wang, Qingsong Lv, Xiaohan Lan, and Yu Zhang. Cross-lingual knowledge graph alignment via graph convolutional networks. In *EMNLP*, pages 349–357, 2018.
19. Yuting Wu, Xiao Liu, Yansong Feng, Zheng Wang, Rui Yan, and Dongyan Zhao. Relation-aware entity alignment for heterogeneous knowledge graphs. In *IJCAI*, pages 5278–5284, 2019.
20. Meng Qu, Jian Tang, and Yoshua Bengio. Weakly-supervised knowledge graph alignment with adversarial learning. *CoRR*, abs/1907.03179, 2019.
21. Fuzhen He, Zhixu Li, Qiang Yang, An Liu, Guanfeng Liu, Pengpeng Zhao, Lei Zhao, Min Zhang, and Zhigang Chen. Unsupervised entity alignment using attribute triples and relation triples. In *DASFAA*, pages 367–382, 2019.
22. Fabian M. Suchanek, Serge Abiteboul, and Pierre Senellart. PARIS: probabilistic alignment of relations, instances, and schema. *PVLDB*, 5(3):157–168, 2011.
23. Vladimir I Levenshtein. Binary codes capable of correcting deletions, insertions, and reversals. In *Soviet physics doklady*, volume 10, pages 707–710, 1966.
24. Bora Edizel, Aleksandra Piktus, Piotr Bojanowski, Rui Ferreira, Edouard Grave, and Fabrizio Silvestri. Misspelling oblivious word embeddings. In *NAACL-HLT*, pages 3226–3234, 2019.
25. Xinyan Dai, Xiao Yan, Kaiwen Zhou, Yuxuan Wang, Han Yang, and James Cheng. Convolutional embedding for edit distance. In *SIGIR*, pages 599–608, 2020.
26. Xin Mao, Wenting Wang, Huimin Xu, Man Lan, and Yuanbin Wu. MRAEA: an efficient and robust entity alignment approach for cross-lingual knowledge graph. In *WSDM*, pages 420–428, 2020.
27. Piotr Bojanowski, Edouard Grave, Armand Joulin, and Tomas Mikolov. Enriching word vectors with subword information. *Transactions of the Association for Computational Linguistics*, 5:135–146, 2017.
28. Hao Nie, Xianpei Han, Le Sun, Chi Man Wong, Qiang Chen, Suhui Wu, and Wei Zhang. Global structure and local semantics-preserved embeddings for entity alignment. In *IJCAI*, pages 3658–3664, 2020.

29. Jinzhu Yang, Wei Zhou, Lingwei Wei, Junyu Lin, Jizhong Han, and Songlin Hu. RE-GCN: relation enhanced graph convolutional network for entity alignment in heterogeneous knowledge graphs. In *DASFAA*, pages 432–447, 2020.
30. Xin Mao, Wenting Wang, Huimin Xu, Yuanbin Wu, and Man Lan. Relational reflection entity alignment. In *CIKM*, pages 1095–1104, 2020.

Chapter 9
Multimodal Entity Alignment

Abstract In various tasks related to artificial intelligence, data is often present in multiple forms or modalities. Recently, it has become a popular approach to combine these different forms of information into a knowledge graph, creating a multi-modal knowledge graph (MMKG). However, multi-modal knowledge graphs (MMKGs) often face issues of insufficient data coverage and incompleteness. In order to address this issue, a possible strategy is to incorporate supplemental information from other multi-modal knowledge graphs (MMKGs). To achieve this goal, current methods for aligning entities could be utilized; however, these approaches work within the Euclidean space, and the resulting entity representations can distort the hierarchical structure of the knowledge graph. Additionally, the potential benefits of visual information have not been fully utilized.

To address these concerns, we present a new approach for aligning entities across multiple modalities, which we call hyperbolic multi-modal entity alignment (HMEA). This method expands upon the conventional Euclidean representation by incorporating a hyperboloid manifold. Initially, we utilize hyperbolic graph convolutional networks(HGCN) to acquire structural representations of entities. In terms of visual data, we create image embeddings using the densenet model and subsequently map them into the hyperbolic space utilizing HGCN. Lastly, we merge the structural and visual representations within the hyperbolic space and utilize the combined embeddings to forecast potential entity alignment outcomes. Through a series of thorough experiments and ablation studies, we validate the efficacy of our proposed model and its individual components.

9.1 Introduction

In recent times, there has been a noticeable trend of integrating multimedia data into knowledge graphs (KGs) to facilitate cross-modal activities that involve the interplay of information across multiple modalities, e.g., image and video retrieval [27], video summaries [19], visual entity disambiguation [17], visual question answering [32], etc. To this end, several multi-modal KGs (MMKGs) [16, 28] have been constructed very recently. An example of MMKG can be found in

© The Author(s) 2023
X. Zhao et al., *Entity Alignment*, Big Data Management,
https://doi.org/10.1007/978-981-99-4250-3_9

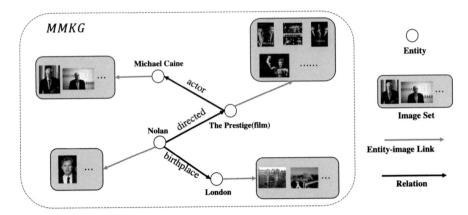

Fig. 9.1 An example of MMKG

Fig. 9.1. For this study, we focus on MMKGs that consist of two modalities, namely, the KG structural details and visual information, while retaining a generalizable approach.

Example Figure 9.1 shows a partial MMKG, which consists of entities, image sets, and the links between them. To elaborate, the KG structural data entails the relationships between the different entities, whereas the visual data is sourced from the sets of images. For the entity `The Prestige`, its image set may contain scenes, actors, posters, etc.

However, many of the current MMKGs have been sourced from restricted data sources, causing them to have inadequate domain coverage [22]. To broaden the scope of these MMKGs, one potential solution is to incorporate valuable knowledge from other MMKGs. An essential step in consolidating knowledge across MMKGs is to identify matching entities in different KGs, given that entities serve as the links that connect the diverse KGs. This technique is also referred to as multi-modal entity alignment (MMEA).

MMEA is a complex undertaking that necessitates the modeling and amalgamation of information from multiple modalities. For the *KG structural information*, existing entity alignment (EA) approaches [3, 9, 25, 33] can be directly adopted to generate entity structural embeddings for MMEA. These methods usually utilize TransE-based or graph convolutional network(GCN)-based models [1, 12] to learn entity representations of individual KGs, which are then unified using the seed entity pairs. Despite this, all of these techniques generate entity representations in the Euclidean space, which can result in significant distortion when embedding real-world graphs that possess scale-free or hierarchical structures [4, 23]. Concerning the *visual information*, the VGG16 model has been utilized to create embeddings for images linked to entities and subsequently employed for alignment. However, the VGG16 model is not adept at extracting valuable features from images, which limits the efficacy of the alignment process. Lastly, the integration of information from both modalities must be executed meticulously to enhance overall effectiveness.

To tackle the problems mentioned above, we introduce a multi-modal entity alignment technique that works in hyperbolic space (HMEA). More specifically, we expand the Euclidean representation to the hyperboloid manifold and utilize the hyperbolic graph convolutional networks (HGCN) to develop structural representations of entities. With regard to visual data, we create image embeddings using the densenet model and also map them into the hyperbolic space with HGCN. Ultimately, we combine the structural embeddings and image embeddings in the hyperbolic space to forecast potential alignments.

To sum up, the key contributions of our technique can be outlined as follows:

- We propose a novel MMEA approach, HMEA, which models and integrates multi-modal information in the hyperbolic space.
- We apply the hyperbolic graph convolutional networks (HGCNs) to develop structural representations of entities and showcase the benefits of the hyperbolic space for knowledge graph representations.
- We use a superior image embedding model to acquire improved visual representations for alignment.
- We perform thorough experimental evaluations to confirm the efficacy of our proposed model.

Organization Section 9.2 overviews related work, and the preliminaries are introduced in Sect. 9.3. Section 9.4 describes our proposed approach. Section 9.5 presents experimental results, followed by conclusion in Sect. 9.6.

9.2 Related Work

In this section, we introduce some efforts that are relevant to this work.

9.2.1 Multi-Modal Knowledge Graph

Many knowledge graph construction studies concentrate on organizing and discovering textual data in a structured format, neglecting other resources available on the Web [28]. Nevertheless, real-world applications require cross-modal data, such as image and video retrieval, visual question answering, video summaries, visual commonsense reasoning, and so on. Consequently, multi-modal knowledge graphs (MMKGs) have been introduced, which comprise diverse information (e.g., image, text, KG) and cross-modal relationships. However, building MMKGs poses several challenges. Collecting substantial multi-modal data from search engines is a time-consuming and laborious task. Additionally, MMKGs often have low domain coverage and are incomplete. Integrating multi-modal knowledge from other MMKGs is an effective way to enhance their completeness. Currently, there are few studies about merging different MMKGs. Liu et al. [16] built two pairs of MMKGs and extracted relational, latent, numerical, and visual features for predicting the *SameAs* link between entities. And some approaches of multi-modal knowledge representation involve visual features from entity images for knowledge representation learning; IKRL [31] integrates image representations into an aggregated image-based representation via an attention-based method.

9.2.2 *Representation Learning in Hyperbolic Space*

Essentially, most of the existing GCN models are designed for graphs in Euclidean spaces [2]. However, research has found that graph data exhibits a non-Euclidean structure [18], and embedding real-world graphs with a scale-free or hierarchical structure results in significant distortion [4, 23]. Moreover, recent studies in network science have shown that hyperbolic geometry is ideal for modeling complex networks, as the hyperbolic space can naturally reflect some graph properties [14]. One of the key features of hyperbolic spaces is that they expand more rapidly than Euclidean spaces, which expands exponentially rather than polynomially. Due to the advantages of hyperbolic space in representing graph structure data, there has been growing interest in representation learning in hyperbolic spaces, particularly in learning the hierarchical representation of a graph [20]. Furthermore, Nickel et al. [21] have demonstrated that the Lorentz model of hyperbolic geometry has favorable properties for stochastic optimization and leads to substantially enhanced embeddings, particularly in low dimensions. Additionally, some researchers have begun to extend deep learning methods to hyperbolic space, achieving state-of-the-art performance on link prediction and node classification tasks [7, 8, 26].

9.3 Preliminaries

In this section, we start by providing a formal definition of the MMEA task. Then, we provide a brief overview of the GCN model. Lastly, we introduce the fundamental principles of hyperbolic geometry, which serve as the foundation for our proposed model.

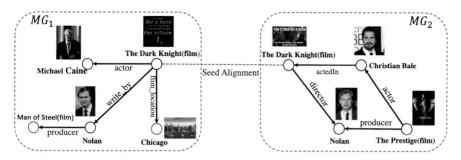

Fig. 9.2 An example of MMEA. Seed entity pairs are connected by dashed lines. For clarity, we only choose an image to represent the set of images of an entity

9.3.1 Task Formulation

The goal of MMEA is to align entities in two MMKGs. An MMKG typically encompasses information in several modalities. In this study, we concentrate on the KG structural information and visual information, without any loss of generality. Formally, we represent MMKGs as $MG = (E, R, T, I)$, where E, R, T, and I denote the sets of entities, relations, triples, and images, respectively. A relational triple $t \in T$ can be represented as (e_1, r, e_2), where $e_1, e_2 \in E$ and $r \in R$. An entity e is associated with multiple images $I_e = \{i_e^0, i_e^1, \ldots, i_e^n\}$.

Given two MMKGs, $MG_1 = (E_1, R_1, T_1, I_1)$, $MG_2 = (E_2, R_2, T_2, I_2)$, and seed entity pairs (pre-aligned entity pairs for training) $S = \{(e_s^1, e_s^2) | e_s^1 \leftrightarrow e_s^2, e_s^1 \in E_1, e_s^2 \in E_2\}$, where $\leftrightarrow$ represents equivalence, the task of MMEA can be defined as discovering more aligned entity pairs $\{(e^1, e^2) | e^1 \in E_1, e^2 \in E_2\}$. We use the following example to further illustrate this task.

Example Figure 9.2 shows two partial MMKGs. The equivalence between The Dark Knight in MG_1 and The Dark Knight in MG_2 is known in advance. EA aims to detect potential equivalent entity pairs, e.g., Nolan in MG_1 and Nolan in MG_2, using the known alignments. □

9.3.2 Graph Convolutional Neural Networks

GCNs [10, 13] are a neural network type that works directly with graph data. A GCN model comprises several stacked GCN layers. The inputs to the l-th layer of the GCN model are node feature vectors and the graph's structure. $H^{(l)} \in R^{n \times d^l}$ is a vertex feature representation, where n is the number of vertices and d^l is the dimensionality of feature matrix. $\hat{A} = D^{-\frac{1}{2}}(A + I)D^{-\frac{1}{2}}$ represents the symmetric normalized adjacency matrix. The identity matrix I is added to the adjacency matrix A to obtain self-loops for each node, and the degree matrix $D = \sum_j(A_{ij}+I_{ij})$. The output of the l-th layer is a new feature matrix $H^{(l+1)}$ by the following convolutional computation:

$$H^{(l+1)} = \sigma(\hat{A}H^{(l)}W^{(l)}). \tag{9.1}$$

9.3.3 Hyperboloid Manifold

We provide a brief overview of the critical concepts in hyperbolic geometry. For a more comprehensive description, please refer to [6]. Hyperbolic geometry refers to a non-Euclidean geometry that features a constant negative curvature used to measure how a geometric object differs from a flat plane. In this work, we use the d-dimensional Poincare ball model with negative curvature $-c$ ($c > 0$): $P^{(d,c)} = \{x \in R^d : \|x\|^2 < \frac{1}{c}\}$, where $\| \cdot \|$ is the L_2 norm. For each point $x \in P^{(d,c)}$, the tangent space T_x^c is a d-dimensional vector space at point x, which contains all possible directions of paths in $P^{(d,c)}$ leaving from x. Next, we present several fundamental actions in the hyperbolic space, which play a critical role in our proposed model.

Exponential and Logarithmic Maps Specifically, let v be the feature vector in the tangent space T_0^c; o is a point in the hyperbolic space $P^{(d,c)}$, which is also used as a reference point. Let o be the origin, $o = 0$. The tangent space T_0^c can be mapped to $P^{(d,c)}$ via the exponential map:

$$\exp_o^c(v) = \tanh(\sqrt{c}\|v\|)\frac{v}{\sqrt{c}\|v\|}. \tag{9.2}$$

And conversely, the logarithmic map which maps $P^{(d,c)}$ to T_0^c is defined as:

$$\log_o^c(y) = \operatorname{arctanh}(\sqrt{c}\|y\|)\frac{y}{\sqrt{c}\|y\|}. \tag{9.3}$$

Möbius Addition Vector addition does not have a well-defined meaning in the hyperbolic space. Adding the vectors of two points directly, as in Euclidean space, in the Poincare ball could yield a point outside the ball. In this case, the Möbius addition [7] provides an analogue to the Euclidean addition in the hyperbolic space. Here, $\oplus_c$ represents the Möbius addition as:

$$h_i \oplus_c h_j = \frac{\left(1 + 2c\langle h_i, h_j\rangle + c\|h_j\|^2\right)h_i + \left(1 - c\|h_i\|^2\right)h_j}{1 + 2c\langle h_i, h_j\rangle + c^2\|h_i\|^2\|h_j\|^2}. \tag{9.4}$$

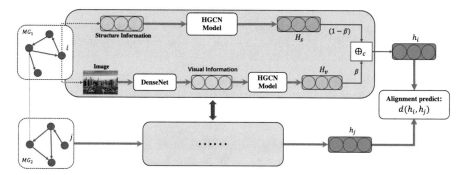

Fig. 9.3 The framework of our proposed method

9.4 Methodology

In this section, we present our proposed approach HMEA, which operates in the hyperbolic space. The framework is shown in Fig. 9.3. We first adopt HGCN to obtain the structural embeddings of entities. Subsequently, we transform the corresponding entity images into visual embeddings employing the densenet model, which are further projected into the hyperbolic space. In the end, we join these embeddings in the hyperbolic space and predict the alignment outcomes utilizing a pre-determined hyperbolic distance. We use the following example to illustrate our proposed model.

> *Example* Further to the previous example, by using structural information, it is easy to detect that Nolan in MG_1 is equivalent to Nolan in MG_2. However, solely relying on structural data is insufficient and might result in an incorrect alignment of Michael Caine in MG_1 with Christian Bale in MG_2. In this scenario, the utilization of visual information would be highly beneficial as the images of Michael Caine in MG_1 and Christian Bale in MG_2 are significantly dissimilar. Consequently, we consider both structural and visual information for alignment. □

In the following, we elaborate on the various components of our proposal.

9.4.1 Structural Representation Learning

We acquire the structural representation of MMKGs by employing hyperbolic graph convolutional neural networks, which extends convolutional computation to manifold space and leverages the effectiveness of both graph neural networks and hyperbolic embeddings. Initially, we transform the input Euclidean features to the hyperboloid manifold. Then, through *feature transformation*, *message passing*, and *nonlinear activation* in the hyperbolic space, we can get the hyperbolic structural representations.

Mapping Input Features to Hyperboloid Manifold In general, the input node features are produced by pre-trained Euclidean neural networks, and hence, they exist in the Euclidean space. We begin by establishing a conversion from Euclidean features to the hyperbolic space.

Here, we assume that the input Euclidean features $x^E \in T_{\mathbf{o}} H_c$, where $T_{\mathbf{o}} H_c$ represent the tangent space referring to $\mathbf{o}$, and $\mathbf{o} \in H_c$ denotes the north pole (origin) in hyperbolic space. We obtain the hyperbolic feature matrix x^H via: $x^H = \exp_o^c(x^E)$, where $\exp_o^c(\cdot)$ is defined in Eq. (9.2).

Feature Transformation and Propagation The core operations in hyperbolic structural learning, similar to GCN, are feature transformation and message passing. While these operations are well-established in the Euclidean space, they are considerably more complex in the hyperboloid manifold. One possible solution is to perform these functions with trainable parameters in the *tangent space* of a point within the hyperboloid manifold, as the tangent space is Euclidean. To this end, we utilize the $\exp(\cdot)$ map and $\log(\cdot)$ map to convert between the hyperboloid manifold and the tangent space. This enables us to make use of the tangent space $T_{\mathbf{o}} H_c^d$ for executing Euclidean operations.

The initial step involves using the logarithmic map to map the hyperbolic representation $x_v^H \in R^{1 \times d}$ of node v to the tangent space $T_{\mathbf{o}} H_c^d$. Next, in $T_{\mathbf{o}} H_c^d$, we compute the feature transformation and propagation rule for node v as:

$$x_v^T = \hat{A} \log_{\mathbf{o}}^c \left(x_v^H \right) W, \tag{9.5}$$

where $x_v^T \in R^{1 \times d'}$ denotes the feature representation in the tangent space and $\hat{A}$ represents the symmetric normalized adjacency matrix; W is a $d' \times d$ trainable weight matrix.

Nonlinear Activation with Different Curvatures Once the features have been transformed in the tangent space, a nonlinear activation function $\sigma^{\otimes^{c_l, c_{l+1}}}$ is applied to learn nonlinear transformations. Specifically, in the tangent space $T_{\mathbf{o}} H_{c_l}^d$ of layer l, Euclidean nonlinear activation is performed before mapping the features to the manifold of the next layer:

$$\sigma^{\otimes^{c_l, c_{l+1}}} \left(x_v^T \right) = \exp_{\mathbf{o}}^{c_{l+1}} \left(\sigma \left(\log_{\mathbf{o}}^{c_l} \left(x_v^T \right) \right) \right), \tag{9.6}$$

where the hyperbolic curvatures at layer l and $l+1$ are denoted as $-1/c_l$ and $-1/c_{l+1}$, respectively. The activation function σ used is the ReLU($\cdot$) function. This step is critical in enabling us to vary the curvature smoothly at each layer, which is necessary for achieving good performance due to limitations in machine precision and normalization.

Based on the hyperboloid feature transformation and nonlinear activation, the convolutional computation in the hyperbolic space is redefined as:

$$H^{l+1} = \exp_{\mathbf{o}}^{c_{l+1}} \left(\sigma \left(\hat{A} \log_{\mathbf{o}}^{c_l} \left(H^l \right) W \right) \right), \tag{9.7}$$

where the convolutional computation in hyperbolic space involves using learned node embeddings in the hyperbolic space at layer $l + 1$ and layer l, represented respectively as $H^{l+1} \in R^{n \times d^{l+1}}$ and $H^l \in R^{n \times d^l}$. The initial embeddings are represented as $H^0 = x^H$. The symmetric normalized adjacency matrix is represented by $\hat{A}$, and the trainable weight matrix is represented by W, which has dimensions $d^l \times d^{l+1}$.

9.4.2 Visual Representation Learning

The densenet model [11] is used to learn image embeddings, which has been pre-trained on the ImageNet dataset [5]. The softmax layer in densenet is removed and 1920-dimensional embeddings are obtained for all images in the MMKGs. These embeddings are then projected into the hyperbolic space using HGCN to enhance their expressive power.

9.4.3 Multi-Modal Information Fusion

As both visual and structural information can impact the alignment results. To combine these two types of information, we propose a novel method that merges the *structural information* and *visual information* of MMKGs. Specifically, we obtain the merged representation of entity e_i in the hyperbolic space using the following approach:

$$h_i = \left(\beta \cdot H_s^i \right) \oplus_c \left((1 - \beta) \cdot H_v^i \right), \tag{9.8}$$

where H_s and H_v are structural and visual embeddings learned from HGCN model, respectively; the hyper-parameter β is used to adjust the relative weight of the structure and visual features in the final merged representation. The Möbius addition operator $\oplus_c$ is used to combine the structural and visual embeddings. However, the dimensions of the structural and visual representations should be identical.

9.4.4 Alignment Prediction

To predict the alignment results, we compute the distance between the entity representations from two MMKGs. The Euclidean distance and Manhattan distance are popular distance measures used in the Euclidean space [15, 30]. However, in the hyperbolic space, we must use the hyperbolic distance between nodes as the distance measure. For entities e_i in MG_1 and e_j in MG_2, the distance is defined as:

$$d_c \left(\boldsymbol{h}_i, \boldsymbol{h}_j \right) = ||(-\boldsymbol{h}_i) \oplus_c \boldsymbol{h}_j||, \qquad (9.9)$$

where $\boldsymbol{h}_i$ and $\boldsymbol{h}_j$ denote the merged embeddings of e_i and e_j in the hyperbolic space, respectively; $|| \cdot ||$ is the L_1 norm; the operator $\oplus_c$ is the Möbius addition.

We expect the distance to be small for equivalent entities and large for nonequivalent ones. To align a specific entity e_i in MG_1, our approach calculates the distances between e_i and all entities in MG_2 and presents a ranked list of entities as candidate alignments.

9.4.5 Model Training

To embed equivalent entities as closely as possible in the vector space, we utilize a set of established entity alignments (known as seed entities) S as training data to train the model. Specifically, we minimize the margin-based ranking loss function during model training:

$$L = \sum_{(e,v) \in S} \sum_{(e',v') \in S'_{(e,v)}} [d_c \left(\boldsymbol{h}_e, \boldsymbol{h}_v \right) + \gamma - d_c \left(\boldsymbol{h}_{e'}, \boldsymbol{h}_{v'} \right)]_+ \quad , \qquad (9.10)$$

where $[x]_+ = \max\{0, x\}$; (e, v) represents a seed entity pair and S is the set of entity pairs; $S'_{(e,v)}$ represents the set of negative instances created by altering (e, v), i.e., by substituting e or v with a randomly selected entity from either MG_1 or MG_2; $\gamma > 0$ denotes the margin hyper-parameter that separates positive and negative instances. The margin-based loss function stipulates that the distance between entities in positive pairs should be small, and the distance between entities in negative pairs should be large.

9.5 Experiment

9.5.1 Dataset and Evaluation Metric

In this study, we utilized datasets sourced from FreeBase, DBpedia, and YAGO, which were created by Liu et al. [16]. These datasets were developed by starting with FB15K to establish multi-modal knowledge graphs, which were then aligned with entities from other knowledge graphs such as DB15K and YAGO15K through reference links. Our experiments focused on two pairs of multi-modal knowledge graphs: FB15K-DB15K and FB15K-YAGO15K.

Due to the absence of original images in the datasets, we acquired the corresponding images for each entity using the URIs provided in [17]. To achieve this, we developed a Web crawler that can extract query results from image search engines, i.e., Google Images,[1] Bing Images,[2] and Yahoo Image Search.[3] Following this, we allocated the images obtained from various search engines to different MMKGs, thereby showcasing the dissimilarity among different MMKGs.

The detailed information on the datasets is provided in Table 9.1. Each dataset comprises approximately 15,000 entities and over 11,000 sets of entity images. The *Images* column represents the number of entities that possess the image sets. These alignments are given by the *SameAs* predicates that have been previously found. In the experiments, the known equivalent entity pairs are used for model training and testing.

Evaluation Metric We utilize $Hits@k$ as the evaluation metric to gauge the efficacy of all the approaches. This metric determines the percentage of correctly aligned entities that are ranked among the top-k candidates.

9.5.2 Experimental Setting and Competing Approaches

Experimental Setting To analyze the effectiveness of the methods across various percentages of the provided alignments $P(\%)$, we evaluate the methods with low

Table 9.1 Statistic of the MMKG datasets

Datasets	Entities	Relations	Rel.Triples	Images	SameAs
FB15K	14,951	1,345	592,213	13,444	
DB15K	14,777	279	99,028	12,841	12,846
YAGO15K	15,404	32	122,886	11,194	11,199

---

[1] https://www.google.com/imghp?hl=EN.

[2] https://www.bing.com/image.

[3] https://images.search.yahoo.com/.

(20%), medium (50%), and high percentage (80%) of the given seed entity pairs. The remaining *sameAs* triples are used for test. To ensure fairness, we have maintained the same number of dimensions (i.e., 400) for both GCN-Align and HMEA. The other parameters of GCN-Align follow [29]. For the parameters of our approach HMEA, we have created six negative samples for each positive sample. The margin hyper-parameters used in the loss function are $\gamma_{HMEA-s} = 0.5$ and $\gamma_{HMEA-v} = 1.5$, respectively. We optimized HMEA using the Adam optimizer.

Competing Approaches To showcase the effectiveness of our proposed model, we have selected three state-of-the-art approaches as competitors:

- GCN-Align [29] utilizes GCN to encode the structural information of entities and then combines relation and image embeddings for the purpose of entity alignment.
- PoE [16] is based on the product of expert model. It computes the scores of facts under each modality and learns the entity embeddings for entity alignment. PoE combines information from two modalities. Additionally, we compare our approach with the PoE-s variant, which solely utilizes the structural information.
- IKRL [31] integrates image representations into an aggregated image-based representation via an attention-based method. The method was initially proposed in the domain of knowledge representation, and we adapted it to address the MMEA problem.

In order to showcase the advantages of hyperbolic geometry, particularly in the learning of structural features, we have conducted preliminary experiments which solely utilize the *structural information* for EA, resulting in HMEA-s, GCN-Align-s, and PoE-s. In addition, to evaluate the contribution of *visual information*, we compare PoE, GCN-Align, and HMEA with just *visual information*, namely, PoE-v, GCN-Align-v, and HMEA-v.

9.5.3 Results

Table 9.2 displays the results, indicating that HMEA exhibits the most superior performance in all scenarios. Notably, in the case of FB15K-YAGO15K with 80% seed entity pairs, HMEA outperforms PoE and GCN-Align by almost 15% in terms of $Hits@1$. With 20% seed entity pairs, our approach also shows better results and the improvement of $Hits@1$ is around 2% and $Hits@10$ is up to 20%. Based on the results obtained from PoE, it is evident that there is only a slight improvement in performance from $Hits@1$ to $Hits@10$, with the range being between 4 and 9%. In contrast, the enhancements in performance from $Hits@1$ to $Hits@10$ observed for HMEA are at least 20% across all scenarios. Moreover, it is worth noting that HMEA achieves significantly better results than IKRL.

Table 9.3 demonstrates that even when utilizing solely *structural information*, HMEA-s still achieves superior results compared to the other two methods. Specifically, our proposed approach outperforms GCN-Align-s by almost 5% in terms of

Table 9.2 Alignment prediction on both datasets for different percentages of P

FB15K-DB15K	20%		50%		80%	
	Hits@1	*Hits*@10	*Hits*@1	*Hits*@10	*Hits*@1	*Hits*@10
PoE	11.1	17.8	23.5	33.0	34.4	40.6
GCN-Align	5.35	17.11	13.85	34.31	22.18	48.95
IKRL	1.01	2.40	2.77	5.79	5.41	11.09
HMEA	**12.65**	**36.86**	**26.23**	**58.08**	**41.68**	**78.55**
FB15K-YAGO15K	20%		50%		80%	
	Hits@1	*Hits*@10	*Hits*@1	*Hits*@10	*Hits*@1	*Hits*@10
PoE	8.7	13.3	18.5	24.7	28.9	34.3
GCN-Align	6.76	17.99	16.47	35.85	28.75	53.05
IKRL	0.86	1.75	1.95	3.73	3.57	7.14
HMEA	**10.51**	**31.27**	**26.50**	**58.08**	**43.30**	**80.11**

The bold figures in the tables represent the best result

Table 9.3 Results of three methods with *structural information*

FB15K-DB15K	20%		50%		80%	
	Hits@1	*Hits*@10	*Hits*@1	*Hits*@10	*Hits*@1	*Hits*@10
PoE-s	10.7	16.5	22.9	31.7	33.6	38.6
GCN-Align-s	5.35	17.11	13.85	34.31	22.18	48.95
HMEA-s	**11.73**	**33.56**	**24.84**	**56.69**	**40.87**	**76.77**
FB15K-YAGO15K	20%		50%		80%	
	Hits@1	*Hits*@10	*Hits*@1	*Hits*@10	*Hits*@1	*Hits*@10
PoE-s	8.4	12.3	18.0	23.1	28.1	31.9
GCN-Align-s	6.76	17.99	16.47	35.85	28.75	53.05
HMEA-s	**9.66**	**28.96**	**25.37**	**56.60**	**42.63**	**78.42**

The bold figures in the tables represent the best result

Hits@1 on FB15K-DB15K and by 3% on FB15K-YAGO15K with 20% seed alignments. When using 50 and 80% seed entity pairs, HMEA-s shows significant improvements in performance. The improvements range from 10 to 18% regarding *Hits*@1 and from 20 to 30% in terms of *Hits*@10. These results suggest that our approach excels in capturing precise hierarchical structural representations.

Table 9.4 presents the results when incorporating *visual information* into the model. We compare the performance of three variants: PoE-v, GCN-Align-v, and HMEA-v. The results indicate that GCN-Align-v does not produce valuable visual representations for MMEA. However, even when utilizing only structural information, HMEA-v still achieves better results than PoE-v. Specifically, our proposed approach outperforms PoE-v slightly in both datasets for *Hits*@1, by less than 1% with 20% seed alignments. On FB15K-DB15K dataset, when using 80% seeds, our proposed approach HMEA-v demonstrates significant improvements in performance. The improvements are around 7% regarding *Hits*@1 and 18% in terms of *Hits*@10. These results indicate that our proposed method is effective

Table 9.4 Comparison of three methods with *visual information*

FB15K-DB15K	20%		50%		80%	
	Hits@1	*Hits@10*	*Hits@1*	*Hits@10*	*Hits@1*	*Hits@10*
PoE-v	0.8	2.7	1.3	3.8	1.7	5.9
GCN-Align-v	0.0	0.0	0.0	0.0	0.0	0.0
HMEA-v	**1.77**	**8.08**	**3.33**	**12.65**	**9.05**	**24.20**
FB15K-YAGO15K	20%		50%		80%	
	Hits@1	*Hits@10*	*Hits@1*	*Hits@10*	*Hits@1*	*Hits@10*
PoE-v	0.7	2.4	1.1	3.2	1.7	5.5
GCN-Align-v	0.0	0.0	0.0	0.0	0.0	0.0
HMEA-v	**1.35**	**5.43**	**2.71**	**11.15**	**5.79**	**18.07**

The bold figures in the tables represent the best result

in learning visual features and incorporating them into the model to improve the overall performance.

9.5.4   Ablation Experiment

In this work, we consider multiple modalities of information in MMKGs. Specifically, we take into account the structural and visual aspects of the information. To further confirm the usefulness of multi-modal knowledge for MMEA, we carry out an ablation experiment. In addition, upon comparing HMEA and HMEA-s in Tables 9.2 and 9.3, we observe that incorporating visual information in our approach results in slightly better performance. The improvements are approximately 1% in terms of $Hits@1$. Moreover, by comparing HMEA and HMEA-v in Tables 9.2 and 9.4, we can also conclude that the structural information plays a significant role. From the ablation study, we can conclude that MMEA primarily relies on the *structural information*, but the *visual information* still plays a useful role. Furthermore, the study also highlights that the combination of these two types of information leads to even better results.

9.5.5   Case Study

A key property of hyperbolic spaces is their exponential expansion, which means that they expand much faster than Euclidean spaces that expand polynomially. This property can be advantageous for distinguishing between similar entities since the neighbor nodes of a central node can be distributed in a larger space, resulting in greater distances between them.

To demonstrate the effectiveness of hyperbolic embeddings, we conducted a case study using `Michael Caine` as the root node. We visualized the embeddings of 1-hop film-related entities learned from both GCN-Align and HMEA separately, in the PCA-projected spaces shown in Fig. 9.4. We observed that for entities of the same type or with similar structural information, such as entity `Alfie` and `B-o-B`, their Euclidean embeddings (generated via GCN-Align) are placed closely together. In contrast, the distances between such entities in hyperbolic space are relatively farther apart, with only a few exceptions. This validates that the hyperbolic structural representation can help distinguish between similar entities. Furthermore, by placing similar entities (in the same KG) far apart, the hyperbolic representation can facilitate the alignment process across KGs.

An example can be seen in Fig. 9.4a, where entity `Alfie` in FB15K is closest to entity `B-o-B`, which is incorrect. However, in Fig. 9.4b, entity `B-o-B` is placed far away from `Alfie`, and the closest entity to `Alfie` is its equivalent entity in

Fig. 9.4 The embeddings of 1-hop film-related neighbor entities of `Michael Caine` generated from GCN-Align and HMEA separately in the PCA-projected space. The green points represent entities in FB15K; red points represent entities in DB15K. For simplicity, we annotate part of entities. `B-o-B` is the abbreviation of `Battle of Britain`. (**a**) Embedding generated from GCN-Align. (**b**) Embedding generated from HMEA

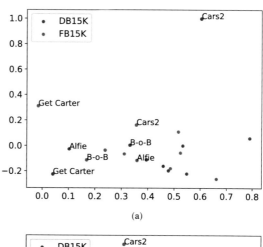

(a)

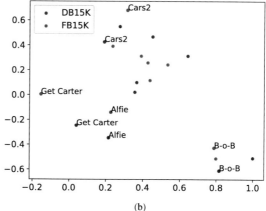

(b)

Table 9.5 Details of the cross-lingual datasets

Datasets		Entities	Relations	Attributes	Rel.triples	Attr.triples
DBP15K$_{ZH-EN}$	Chinese	66,469	2,830	8,113	153,929	379,684
	English	98,125	2,137	7,173	237,674	567,755
DBP15K$_{JA-EN}$	Japanese	65,744	2,043	5,882	164,373	354,619
	English	95,680	2,096	6,066	233,319	497,230
DBP15K$_{FR-EN}$	French	66,858	1,379	4,547	192,191	528,665
	English	105,889	2,209	6,422	278,590	576,543

DB15K. By using hyperbolic projections, similar entities in the same KG are well distinguished and placed far apart, reducing the likelihood of alignment mistakes.

9.5.6 Additional Experiment

The cross-lingual EA datasets are the most commonly used datasets for evaluating EA methods. We included experiments on these datasets to demonstrate that our proposed approach is effective for popular datasets, including the cross-lingual EA task. Note that diverse languages are not taken as multiple modalities, and the cross-lingual EA is in essence single-modal EA. We use the DBP15K datasets in the experiments, which were built by Sun et al. [24]. As shown in Table 9.5, the datasets were generated from DBpedia, which contains rich inter-language links between different language versions of Wikipedia. Each dataset contains data in different languages and 15,000 known inter-language links connecting equivalent entities in two KGs, which are used for model training and testing. Following the setting in [29], we use 30% of inter-language links for training, and 70% of them for testing. $Hits@k$ is used as the evaluation measure.

The dimensions of both structural and attribute embeddings were set to 300 dimensions for GCN-Align. GCN-Align-s and HMEA-s represent adopting structural information; GCN-Align-a and HMEA-a represent adopting attribute information; and GCN-Align and HMEA combine both the structural information and attribute information.

Table 9.6 shows that in all datasets, HMEA-s outperforms GCN-Align-s, with improvements of around 7% in terms of $Hits@1$ and more than 10% in terms of $Hits@10$. These results demonstrate that HMEA benefits from hyperbolic geometry and is able to capture better structural features. Furthermore, our proposed approach achieves better results compared to GCN-Align as it combines both structural and attributive information, resulting in an approximately 10% increase in $Hits@1$. Regarding attribute information, it is worth noting that our approach, HMEA-a, outperforms GCN-Align-a by a significant margin. Specifically, our approach achieves an approximately 15% improvement in $Hits@1$ across all datasets.

Table 9.6 Result in cross-lingual datasets

DBP15K$_{ZH-EN}$	ZH-EN		EN-ZH	
	$Hits@1$	$Hits@10$	$Hits@1$	$Hits@10$
GCN-Align-s	39.42	71.34	33.60	65.23
HMEA-s	**46.23**	**82.36**	**44.53**	**81.95**
GCN-Align-a	13.44	40.94	12.54	38.78
HMEA-a	**33.99**	**71.15**	**32.80**	**69.79**
GCN-Align	43.08	75.92	36.25	69.17
HMEA	**54.04**	**87.88**	**51.88**	**86.57**
DBP15K$_{JA-EN}$	JA-EN		EN-JA	
	$Hits@1$	$Hits@10$	$Hits@1$	$Hits@10$
GCN-Align-s	39.95	72.72	36.09	67.43
HMEA-s	**47.63**	**83.96**	**47.24**	**83.96**
GCN-Align-a	9.27	31.85	8.78	31.89
HMEA-a	**28.36**	**63.99**	**27.73**	**63.97**
GCN-Align	42.51	75.74	38.31	70.49
HMEA	**53.06**	**87.47**	**52.65**	**87.41**
DBP15K$_{FR-EN}$	FR-EN		EN-FR	
	$Hits@1$	$Hits@10$	$Hits@1$	$Hits@10$
GCN-Align-s	38.38	74.45	37.37	71.65
HMEA-s	**44.27**	**83.15**	**43.81**	**83.14**
GCN-Align-a	2.65	13.50	3.02	14.51
HMEA-a	**12.40**	**48.70**	**15.44**	**52.12**
GCN-Align	39.48	76.05	38.44	73.33
HMEA	**48.40**	**86.49**	**48.15**	**86.18**

9.6 Conclusion

This chapter introduces our proposed approach, HMEA, which is a multi-modal EA approach designed to efficiently integrate multi-modal information for EA in MMKGs. To achieve this, our approach extends the Euclidean representation to a hyperboloid manifold and employs HGCN to learn structural embeddings of entities. Additionally, we leverage a more advanced model, densenet, to learn more accurate visual embeddings. These structural and visual embeddings are then aggregated in the hyperbolic space to predict potential alignments. We validate the effectiveness of our proposed approach through comprehensive experimental evaluations. Additionally, we conduct further experiments that confirm the superior performance of HGCN in learning structural features of knowledge graphs in the hyperbolic space.

References

1. A. Bordes, N. Usunier, A. García-Durán, J. Weston, and O. Yakhnenko. Translating embeddings for modeling multi-relational data. In *NIPS*, pages 2787–2795, 2013.
2. S. Cavallari, E. Cambria, H. Cai, K. C.-C. Chang, and V. W. Zheng. Embedding both finite and infinite communities on graphs [application notes]. *IEEE Computational Intelligence Magazine*, 14(3):39–50, 2019.
3. M. Chen, Y. Tian, M. Yang, and C. Zaniolo. Multilingual knowledge graph embeddings for cross-lingual knowledge alignment. *arXiv preprint arXiv:1611.03954*, 2016.
4. W. Chen, W. Fang, G. Hu, and M. W. Mahoney. On the hyperbolicity of small-world and treelike random graphs. *Internet Mathematics*, 9(4):434–491, 2013.
5. J. Deng, W. Dong, R. Socher, L. Li, K. Li, and F. Li. Imagenet: A large-scale hierarchical image database. In *CVPR*, pages 248–255. IEEE Computer Society, 2009.
6. L. P. Eisenhart. *Introduction to differential geometry*. Princeton University Press, 2015.
7. O.-E. Ganea, G. Bécigneul, and T. Hofmann. Hyperbolic entailment cones for learning hierarchical embeddings. *arXiv preprint arXiv:1804.01882*, 2018.
8. C. Gulcehre, M. Denil, M. Malinowski, A. Razavi, R. Pascanu, K. M. Hermann, P. Battaglia, V. Bapst, D. Raposo, A. Santoro, et al. Hyperbolic attention networks. *arXiv preprint arXiv:1805.09786*, 2018.
9. Y. Hao, Y. Zhang, S. He, K. Liu, and J. Zhao. A joint embedding method for entity alignment of knowledge bases. In *CCKS*, pages 3–14. Springer, 2016.
10. M. Henaff, J. Bruna, and Y. LeCun. Deep convolutional networks on graph-structured data. *arXiv preprint arXiv:1506.05163*, 2015.
11. G. Huang, Z. Liu, L. Van Der Maaten, and K. Q. Weinberger. Densely connected convolutional networks. In *CVPR*, pages 4700–4708, 2017.
12. T. N. Kipf and M. Welling. Semi-supervised classification with graph convolutional networks. *CoRR*, abs/1609.02907, 2016.
13. T. N. Kipf and M. Welling. Semi-supervised classification with graph convolutional networks. *arXiv preprint arXiv:1609.02907*, 2016.
14. D. Krioukov, F. Papadopoulos, M. Kitsak, A. Vahdat, and M. Boguná. Hyperbolic geometry of complex networks. *Physical Review E*, 82(3):036106, 2010.
15. C. Li, Y. Cao, L. Hou, J. Shi, J. Li, and T. Chua. Semi-supervised entity alignment via joint knowledge embedding model and cross-graph model. In *EMNLP*, pages 2723–2732. Association for Computational Linguistics, 2019.
16. Y. Liu, H. Li, A. Garcia-Duran, M. Niepert, D. Onoro-Rubio, and D. S. Rosenblum. Mmkg: multi-modal knowledge graphs. In *ESWC*, pages 459–474. Springer, 2019.
17. S. Moon, L. Neves, and V. Carvalho. Multimodal named entity disambiguation for noisy social media posts. In *ACL (Volume 1: Long Papers)*, pages 2000–2008, 2018.
18. A. Muscoloni, J. M. Thomas, S. Ciucci, G. Bianconi, and C. V. Cannistraci. Machine learning meets complex networks via coalescent embedding in the hyperbolic space. *Nature communications*, 8(1):1–19, 2017.
19. D. A. Newman and B. G. Schunck. Generating video summaries for a video using video summary templates, Oct. 17 2017. US Patent 9,792,502.
20. M. Nickel and D. Kiela. Poincaré embeddings for learning hierarchical representations. In *NIPS*, pages 6338–6347, 2017.
21. M. Nickel and D. Kiela. Learning continuous hierarchies in the lorentz model of hyperbolic geometry. *arXiv: Artificial Intelligence*, 2018.
22. H. Paulheim. Knowledge graph refinement: A survey of approaches and evaluation methods. *Semantic web*, 8(3):489–508, 2017.
23. E. Ravasz and A.-L. Barabási. Hierarchical organization in complex networks. *Physical review E*, 67(2):026112, 2003.
24. Z. Sun, W. Hu, and C. Li. Cross-lingual entity alignment via joint attribute-preserving embedding. In *ISWC*, pages 628–644. Springer, 2017.

25. Z. Sun, W. Hu, Q. Zhang, and Y. Qu. Bootstrapping entity alignment with knowledge graph embedding. In *IJCAI*, pages 4396–4402, 2018.
26. H.-N. Tran and E. Cambria. A survey of graph processing on graphics processing units. *The Journal of Supercomputing*, 74(5):2086–2115, 2018.
27. R. C. Veltkamp, H. Burkhardt, and H.-P. Kriegel. *State-of-the-art in content-based image and video retrieval*, volume 22. Springer Science & Business Media, 2013.
28. M. Wang, G. Qi, H. Wang, and Q. Zheng. Richpedia: A comprehensive multi-modal knowledge graph. In *JIST*, pages 130–145. Springer, 2019.
29. Z. Wang, Q. Lv, X. Lan, and Y. Zhang. Cross-lingual knowledge graph alignment via graph convolutional networks. In *EMNLP*, pages 349–357, 2018.
30. Y. Wu, X. Liu, Y. Feng, Z. Wang, and D. Zhao. Neighborhood matching network for entity alignment. In *ACL*, pages 6477–6487. Association for Computational Linguistics, 2020.
31. R. Xie, Z. Liu, H. Luan, and M. Sun. Image-embodied knowledge representation learning. pages 3140–3146, 2017.
32. K. Yi, J. Wu, C. Gan, A. Torralba, P. Kohli, and J. Tenenbaum. Neural-symbolic vqa: Disentangling reasoning from vision and language understanding. In *NIPS*, pages 1031–1042, 2018.
33. W. Zeng, X. Zhao, J. Tang, and X. Lin. Collective entity alignment via adaptive features. In *ICDE*, pages 1870–1873. IEEE, 2020.

Printed in the United States
by Baker & Taylor Publisher Services